N. John DiNardo

Nanoscale Characterization of Surfaces and Interfaces

Distribution:
VCH, P.O. Box 10 1161, D-69451 Weinheim, Federal Republic of Germany
Switzerland: VCH, P.O. Box, CH-4020 Basel, Switzerland
United Kingdom and Ireland: VCH, 8 Wellington Court, Cambridge CB1 1HZ, United Kingdom
USA and Canada: VCH, 220 East 23rd Street, New York, NY 10010-4606, USA
Japan: VCH, Eikow Building, 10-9 Hongo 1-chome, Bunkyo-ku, Tokyo 113, Japan

ISBN 3-527-29247-0

N. John DiNardo

Nanoscale Characterization of Surfaces and Interfaces

Weinheim · New York · Basel · Cambridge · Tokyo

N. John DiNardo
Department of Physics and
Atmospheric Science
Drexel University
Philadelphia, PA 19104
USA

Published jointly by
VCH Verlagsgesellschaft, Weinheim (Federal Republic of Germany)
VCH Publishers, New York, NY (USA)

Editorial Directors: Dr. Peter Gregory, Deborah Hollis, Dr. Ute Anton
Production Manager: Dipl. Wirt.-Ing. (FH) Hans-Jochen Schmitt

Cover Illustration: Adapted from an AFM image of a thermally-treated diamond-like-carbon film (see Preface)

Library of Congress Card No. applied for

A catalogue record for this book is available from the British Library

Die Deutsche Bibliothek – CIP-Einheitsaufnahme
DiNardo, N. John:
Nanoscale characterization of surfaces and interfaces / N. John DiNardo. – Weinheim ; New York ; Basel ; Cambridge ; Tokyo : VCH, 1994
ISBN 3-527-29247-0

Printed on acid-free and low chlorine paper

Composition, Printing and Bookbinding: Konrad Triltsch, Druck- und Verlagsanstalt GmbH, D-97070 Würzburg

Printed in the Federal Republic of Germany

To my wife, Trudy,
and our children, Christopher and Johanna

STM image of the Si(111) surface acquired with a JEOL JSTM-4500VT microscope. This image shows reconstructed 7×7 regions terminated by surface steps which surround a terrace on which a Ag/Si(111)-3×1 reconstructed region has formed. Position-dependent tunneling spectra find that, electronically, the 7×7 regions appear metallic while the 3×1 region exhibits a 1.1 eV energy gap. *(J. Carpinelli and H. H. Weitering, University of Tennessee/Oak Ridge National Laboratory)*

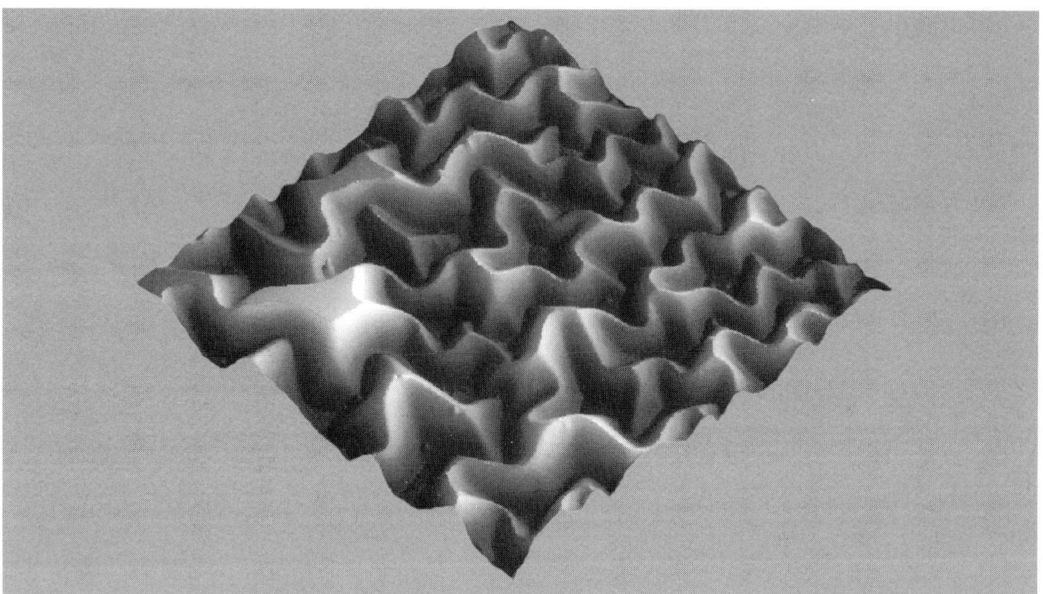

An AFM image of a thermally-treated 500 Å-thick diamond-like-carbon film grown on a Pt substrate by laser ablation using a Digital Instruments Nanoscope III. Thermal processing alters the morphology of the initially flat film by apparently relieving internal stresses, resulting in partial separation of the film from the substrate. *(T. W. Mercer, N. J. DiNardo, Drexel University and Laboratory for Research on the Structure of Matter, University of Pennsylvania; M. P. Siegel, Sandia National Laboratory)*

Preface

The invention of the STM by G. Binnig and H. Rohrer has created an entirely new way for us to view and interact with materials. Imaging materials down to the atomic level with the STM represents a truly remarkable feat of scientific and technological insight. Moreover, the demonstrated capabilities of the STM to actually control materials properties on the atomic and nanometer levels opens new doors to exciting science and technology at the quantum level, which previously could only be considered in textbook equations and gedankenexperiments. The fact that, over the past thirteen years, the STM and related instruments, notably the AFM, have led to a vast number of new developments across a wide array of disciplines clearly indicates the timeliness and usefulness of this class of instrumentation.

This monograph is the product of an extended review article, which I was asked to write for Volume 2 of the series *Materials Science and Technology: A Comprehensive Treatment,* edited by Robert W. Cahn, Peter Haasen, and Edward J. Kramer (Volume Editor Dr. Eric Lifshin). It began with the aim of describing the use of the STM by example, using a host of demonstrated applications. In the light of the rapidly increasing literature on these subjects and related topics, as the manuscript was being written, the work was extended to cover AFM with brief expositions of nanoscale modification and techniques spun-off from STM technology.

In this work, I have attempted to describe a wide array of benchmark experiments and relevant examples and extensions of the capabilities which these made available. I have, therefore, included references to work in physics, chemistry, materials science, biological science, and metrology. The descriptions are by no means exhaustive, but the extensive set of references will hopefully allow the reader to pursue topics of interest in more detail. In addition, considering the explosion of work in the field, several examples of equal importance to those used were not included simply due to lack of space and time. One primary intent is that the reader will gain an appreciation for the diverse fields in which these techniques have established themselves. The photographs on the opposite page are intended to give the reader a taste of the power and beauty of STM and AFM imaging.

I wish to acknowledge the authors whose work has been included in this monograph and their publishers. I am indebted to the editorial and production staff at VCH for their contribution to the task of assembling this monograph. My particular thanks go to Dr. Peter Gregory, Dr. Ute Anton, and Ms. Deborah Hollis on the editoral side and to Wirt.-Ing. Hans-Jochen Schmitt on the production side.

March 1994 N. John DiNardo

Biography

Professor DiNardo is Associate Professor of Physics at Drexel University and Adjunct Associate Professor of Materials Science and Engineering at the University of Pennsylvania. He received his Ph. D. in Physics in 1982 from the University of Pennsylvania. His dissertation involved the use of Electron Energy Loss Spectroscopy (EELS) to study the vibrational structure of adsorbate-metal systems. Subsequently, as a Postdoctoral Fellow at IBM Research (Yorktown Heights), he studied the relationship between vibrational and electronic structure of metal, semiconductor, and polymer interfaces with EELS, Photoelectron Spectroscopy, and Low Energy Electron Diffraction. As a member of the faculty of the Department of Physics and Atmospheric Science at Drexel University since 1984, Professor DiNardo, along with his students, has instituted a program to study the properties of semiconductor surfaces and metal–semiconductor interfaces. A primary area of focus has been the characterization of the semiconductor-to-metal transition using ultrathin alkali-metal films on Si and GaAs surfaces with an array of complementary experimental techniques. Currently, his work extends to vibrational spectroscopy of semiconductor and polymer surfaces, variable-temperature STM/STS of semiconductor interfaces, and the characterization of the geometric and electronic structure of diamond films with STM and AFM.

Nanoscale Characterization of Surfaces and Interfaces

N. John DiNardo

Department of Physics and Atmospheric Science, Drexel University,
Philadelphia, PA, U.S.A.

List of Symbols and Abbreviations

A	constant
A, A_0	vibration amplitude
a, b, c	lattice direction
a_0, b_0, c_0	lattice spacing
$\mathrm{acf}(\boldsymbol{a})$	autocorrelation function
d	sample–tip spacing
d_{31}	piezoelectric coefficient
E, E_T, E_S	energy (tip, sample)
E_F	Fermi energy
E_t	transverse energy
e	electronic charge
F	interatomic force
f	force field
f_r	resonance frequency
h	Planck constant; wall thickness of tube; height
h_0	mean surface height
I	current
I_c	collector current
I_0, I_t	tunneling current
I_t	injected current
j	current density
K_z, $K_{\parallel}$	piezoelectric constants
k, k'	force constants
$\boldsymbol{k}$	wavevector
k_B	Boltzmann constant
k_{bond}	interatomic force constant
M_{TS}	tunneling matrix element between tip wavefunction and sample wavefunction
m	free electron mass
m_a	atomic mass
m_{eff}, m_n, m_t	effective masses
R	tunneling resistance; scattering parameter; radius of curvature of the tip
r	tip radius
s	tip–sample separation
T	absolute temperature
t	time
U	bonding energy
U_{eff}	coulomb energy
V, V_0	voltage
V_s	sample potential
x, y, z	cartesian coordinates
z	tip–sample separation

Δ	corrugation amplitude; energy gap
ϑ_c	critical angle
$\varkappa$	inverse decay length
λ	wavelength
λ_F	Fermi wavelength
μ	chemical potential
ϱ	electrical resistivity
$\varrho(r_0, t)$	charge density
ϱ_m	diameter of orifice set up by a tip atom
ϕ, ϕ_T, ϕ_S	work function
ϕ, ϕ_a	(apparent) barrier height
ψ_S, ψ_T	wavefunction of sample, tip
ψ_S^0, ψ_T^0	surface wavefunction of sample, tip
$\psi_S(z)$, $\psi_T(z)$	wavefunction in vacuum gap
ω	vibrational frequency
ω_0, ω'_0	resonance frequency

AC	alternating current
ADC	analog-to-digital converter
AES	Auger electron spectroscopy
AFM	atomic force microscope; atomic force microscopy
ATP	adenosine 5′-triphosphate
BCS	Bardeen–Cooper–Schrieffer
BEEM	ballistic electron emission microscopy
CBM	conduction band minimum
CDW	charge density wave
CFM	charge force microscope
CITS	current imaging tunneling spectroscopy
DAC	digital-to-analog converter
DAS	dimer-adatom-stacking
DB	domain boundary
DC	direct current
DMPE	dimyristoyl-phosphatidylethanolamine
DNA	deoxyribonucleic acid
DOS	density of states
EELS	electron energy loss spectroscopy
ET	embedded trimer
FIM	field ion microscopy
GIC	graphite intercalation compound
H	honeycomb
HCT	honeycomb-chain-trimer
HM	high resolution optical microscope
HOPG	highly oriented pyrolytic graphite
IETS	inelastic electron tunneling spectroscopy
IPES	inverse photoelectron spectroscopy

JDOS	joint density of states
L–B	Langmuir–Blodgett
LEED	low-energy electron diffraction
LMIS	liquid metal ion source
MFM	magnetic force microscope
ML	monolayers (unit)
NDR	negative differential resistance
OAM	optical absorption microscope
PCM	phase contrast microscope
PDS	photothermal deflection spectroscopy
PMMA	poly(methylmethacrylate)
PSTM	photon scanning tunneling microscope
PTFE	polytetrafluoroethylene
RBS	Rutherford backscattering
RHEED	reflective high-energy electron diffraction
rms	root mean square
SBH	Schottky-barrier height
SBZ	surface Brillouin zone
SCPM	scanning chemical potential microscope
SEM	scanning electron microscope
SICM	scanning ion conductance microscope
SNOM	scanning near-field optical microscope
SPV	surface photovoltage
STEM	scanning transmission electron microscope
STM	scanning tunneling microscope; scanning tunneling microscopy
STP	scanning thermal profiler
STS	scanning tunneling spectroscopy
T	tesla
TEM	transmission electron microscope
TTF-TCNQ	tetrathiafulvalene tetracyanoquinodimethane
UHV	ultra high vacuum
UPD	underpotential deposition
UPS	ultraviolet photoelectron spectroscopy
VBM	valence band maximum
WF	wavefunction
XPS	X-ray photoelectron spectroscopy

1 Introduction

One of the principal objectives in the experimental study of bulk solids is the characterization of atomic structure. Such information is used to provide a link between geometric structure and the other physical properties of a solid. Building upon advances in the experimental and theoretical understanding of the bulk, surfaces and interfaces have been intensively studied in recent years because of their fundamental and technological importance. The surface structure of a single crystal differs from that of the underlying bulk, since truncation at a lattice plane results in the relaxation of the near-surface atomic geometry, often creating unique lateral surface reconstructions. Specific electronic states and vibrational modes can be associated with these structures. The goal of classical surface science is to establish a fundamental atomistic view of surfaces. It is clear that high-technology problems in the current age of materials research rely on such a perspective.

Nanotechnology, i.e., the characterization and manipulation of structures on the nanometer scale, is an increasingly important area of research and development (Crandall and Lewis, 1992). For example, the fabrication of microelectronics devices by epitaxial atom-by-atom growth depends on interactions and energetics governed by phenomena on the atomic scale. Thus, diagnostics using ultrahigh resolution microscopy can help to define the proper conditions under which films may be grown with atomic thickness control. To create structures of nanometer lateral dimensions requires new methods to manipulate atoms and molecules at surfaces. In surface chemistry and catalysis, different crystallographic orientations of a surface or different distributions of defects can alter reactivity by orders of magnitude; the structural and electronic origin of these effects is still under study. Beyond classical surface science, processes at liquid–solid interfaces, e.g. in electrochemical cells, may also be studied at the atomic level. In the new age of structural biology, biological molecules or systems, deposited on a substrate, may likewise be imaged at this level to ascertain critical information on the geometric and electronic structure.

Atomic-scale forces between materials at an interface are important in areas of tribology and adhesion. These are areas where nanoscale properties are directly related to macroscopic phenomena. Likewise, the forces between a sharp tip and a surface can be used for imaging; for example, the repulsive force between a tip and a sample can be exploited to make an atomic scale profilimeter so long as the forces are small enough not to disturb the surface structure. While the size scales of the structures under consideration range from that of individual atoms to tens of nanometers and beyond, it is clear that atomic resolution is not required for all applications.

Surface science has developed as a multidisciplinary field over the past 30 years to study the physical and chemical properties of solid surfaces and interfaces. Until recently, bulk structural techniques, such as transmission electron microscopy or x-ray diffraction, have lacked the sensitivity to view surfaces at atomic resolution, and these techniques still require rather stringent sample preparation procedures and/or experimental set-ups. Many techniques have been developed involving particle scattering or emission to probe the atomic, electronic, and vibrational structure of surfaces since the small mean free path of particles, electrons, ions, or atoms maximizes surface sensitivity. Measurements have typically been performed under high-vacuum or ultrahigh-vacuum conditions in order to preserve surface cleanliness and as a necessary condition to operate particle beam experiments. The major drawback of such spatially-extended probe beam techniques is that the data represents an average over macroscopic areas of the surface (with respect to atomic dimensions) so that the effects of inhomogeneities, which may occur over atomic distances, are often impossible to isolate.

Even high-quality single crystal surfaces are not atomically perfect. Steps, defects and other irregularities may produce important effects which cannot be isolated by averaging over a large area. "Real" surfaces – those used in technological applications – are somewhat removed from the single crystal norm of surface scientists, typically possessing more disorder, defects, crystalline, grains, impurities, etc. It is important to understand the way in which these undesirable structures might undermine the effectiveness of the material for specific applications. Thus, as a complement to the information that can be derived from established techniques, real-space views of surface geometry and spectroscopic capabilities at atomic resolution, can provide a means for comparison with theoretical calculations at the atomic level. The invention of the scanning tunneling microscope (STM) in 1982 (Binnig et al., 1982a, b) made it possible to probe surfaces on the nanometer scale, since the STM can relate geometry and electronic structure at surfaces atom-by-atom. Significantly, the capability of the tunneling probe to operate in liquid and gaseous environments, as well as in vacuum, allows direct analysis of processes at liquid–solid interfaces and of biological structures and processes in vivo. Recognizing that electron tunneling occurs across a small gap, we see that interatomic forces become intimately related to such measurements. This fact has allowed STM technology to be applied to atomic forces measurements at a level of sensitivity of $<10^{-12}$ N by the invention of the atomic force microscope (AFM) (Binnig et al., 1986). Furthermore, the ability to fabricate a microscope probe and control it with nanometer precision has inspired a number of novel allied scanning microprobe techniques (Wickramasinghe, 1990). With the prospect of further device miniaturization on the nanometer scale, the ability of the STM to manipulate atoms and clusters to fabricate microscopic structures and subsequently analyze these structures with the same probe has also been demonstrated, and the microscopic mechanisms for such nanoscale processing are still being studied.

STM possesses a distinct advantage over other microscopy techniques in both vertical and lateral resolution, as seen in Fig. 1. Its application to a vast variety of materi-

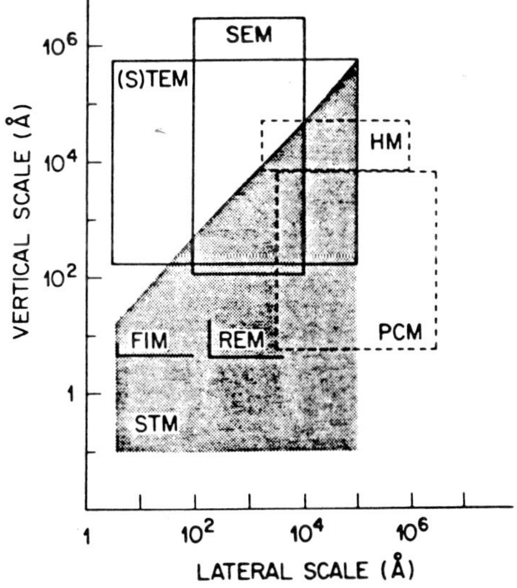

Figure 1. Resolution of various microscopes – HM: high resolution optical microscope; PCM: phase contrast microscope; (S)TEM: (scanning) transmission electron microscope; SEM: scanning electron microscope; FIM: field ion microscope, etc. (Kuk and Silverman, 1989).

als problems and its extension towards the development of allied techniques has occurred over a remarkably short time, and this points to the engineering triumphs which the instrument itself and associated hardware and software represent. Prior to the development of STM, only specialized techniques such as field ion microscopy could provide direct atomic resolution images of surfaces. Atomic resolution with electron microscopy is constrained by sample preparation thereby limiting the range of applicability of the technique. Among other recent developments, real-space imaging of surfaces using electron holography has been reported but limitations still exist for imaging non-periodic structures.

The basic principle of the STM is illustrated in Fig. 2. An atomically-sharp tip, biased with respect to the sample, is positioned at a distance of $\leq 10\,\text{Å}$ from the sample surface. A current in the nanoampere range passes between the tip and the sample due to electron tunneling through the vacuum barrier in the gap region. The current decreases exponentially as the gap distance increases. The tip motion is controlled in three dimensions by piezoelectric transducers that distort by applying a voltage (electric field) across them. A bias of ~ 1 V across a typical piezoelectric element may cause an expansion or concentration of $\sim 10\,\text{Å}$, so that sub-atomic movement of the tip can easily be obtained. It can be assumed that electronic states (orbitals) are localized at each atomic site, so measuring the response of the tip being scanned over the surface can give a picture of the surface atomic structure. The structure may be mapped in the *constant current mode* by recording the feedback-controlled motion of the tip up and down, such that a constant tunneling current is maintained at each x-y position. The structure can also be mapped in the *constant height mode* by recording the modulation of the tunneling current as a function of position, while the tip remains a constant height above the surface. The latter mode is preferred for scanning at high speeds but can only be used on very smooth surfaces. The former is required to obtain topographic images of rough surfaces.

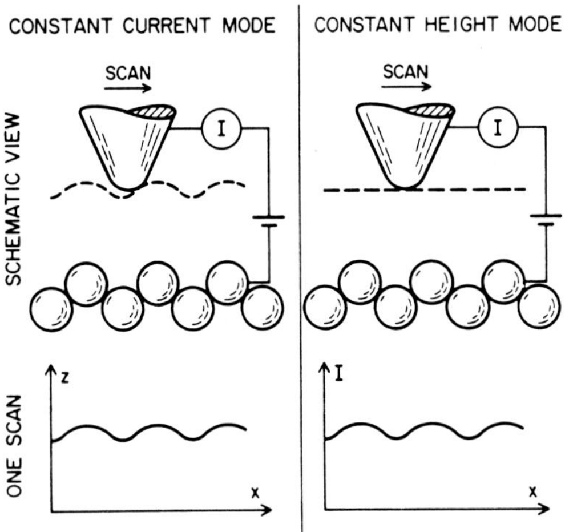

Figure 2. Schematic representation of the operating concept of an STM operating in the *constant current mode* or the *constant height mode* (Hansma and Tersoff, 1987).

Figure 3 shows a schematic diagram of the first STM of Binnig and Rohrer, where the tip is mounted on a piezoelectric tripod assembly, and the sample (S) is mounted on a "louse" (L). The louse is a device which can "walk" the tip on the base by sequences of electrostatic clamping and piezoelectric distortions.

The ability to image the electronic structure is a natural byproduct of STM, and this provides a natural means of obtaining spatially-resolved spectroscopic images. As in tunneling spectroscopy at metal–oxide–metal junctions (Giaver, 1960), electrons pass from filled electronic states (or bands) to unoccupied states (or bands). Depending on the sign of the tunneling bias,

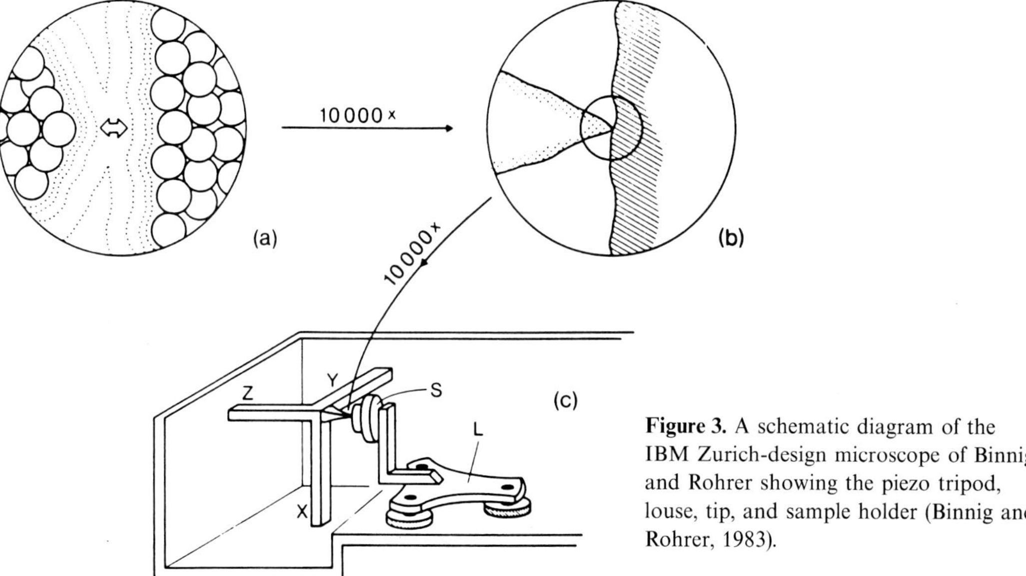

Figure 3. A schematic diagram of the IBM Zurich-design microscope of Binnig and Rohrer showing the piezo tripod, louse, tip, and sample holder (Binnig and Rohrer, 1983).

the current can be directed from sample-to-tip or vice-versa. An illustration of this feature of the STM is given in Fig. 4, which shows two images of the same region of a GaAs(110) surface with biases of opposite polarities (Feenstra et al., 1987b). The positive bias image is displaced from the negative bias image due to the localized atomic orbitals accessible to tunneling. In particular, rehybridization to sp^2-like bonding and relaxation occur at this cleavage plane due to the truncation of the bulk. A filled lone-pair band is localized above As atoms, while an unoccupied band is associated with the Ga atoms. The tunneling current is maximized over As atoms when scanning with a tip bias of +1.9 V, while the current is maximized over Ga atoms with a tip bias of −1.9 V. Stopping at one particular location and scanning the tip bias measures the local (joint) density of states of the sample and tip.

The most general use of the STM is for topographic imaging not necessarily at the atomic level but on length scales from < 10 nm to ≥ 1 μm. Figure 5 shows an application related to mesoscopic physics and an application in practical surface metrology (Denley, 1990). In the first example, an STM was used to measure the surface morphology of quantum dots. As can be seen, the dots are arranged in a periodic

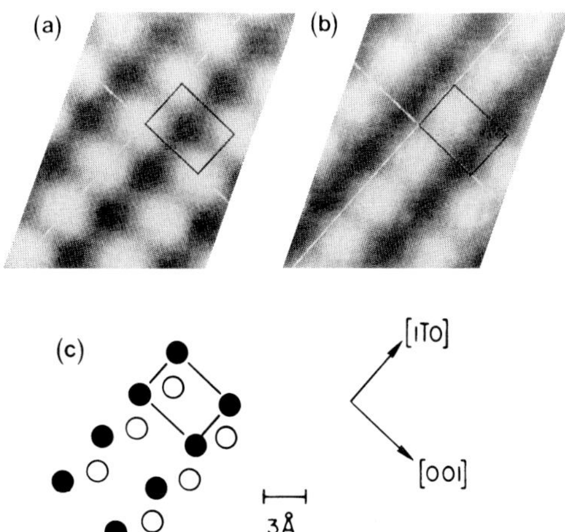

Figure 4. Two views (a, b) of a GaAs(110) surface taken at biases with opposite polarities. With the tip biased positive with respect to the sample, occupied lone-pair states at As sites are imaged; the opposite bias produces a laterally-shifted image of the unoccupied orbitals at Ga sites. (Feenstra et al., 1987b).

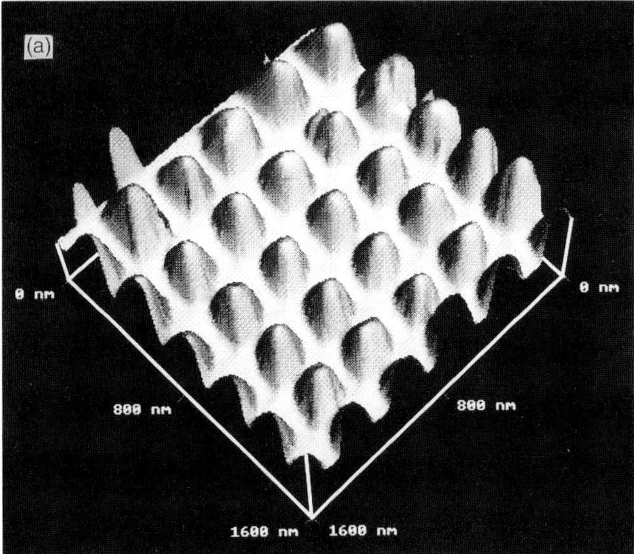

(b)

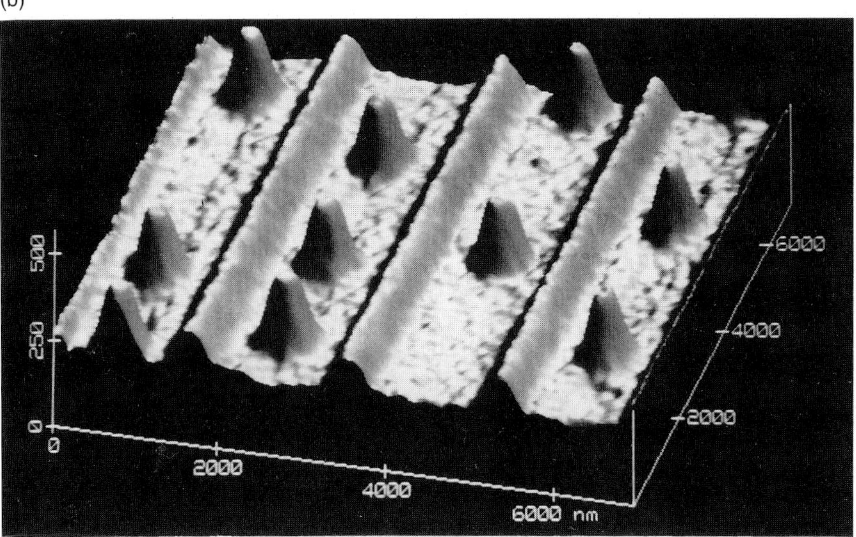

Figure 5. (a) STM image of an array of 100 nm high quantum dots on GaAs; (b) STM image of a compact disc surface (Denley, 1900).

array on GaAs, are each separated by ~300 nm, and are ~100 nm in height. The second example shows the surface of a compact disc over a 6 × 6 μm area.

As an offshoot of STM technology, the atomic force microscope (AFM) provides a means to image surfaces by direct (or proximal) contact of a sharp tip with a surface. Insulating substrates and organic films are more amenable to AFM imaging because a conducting substrate is not required. In an AFM, a sharp tip is mounted on a cantilever and is typically in contact with the surface while the deflection of the cantilever – the force – is monitored optically or by other means. The contact force is typically $<10^{-7}$ N; in constant force operation, the piezoelectrically-actuated sample height modulation in response to hold the cantilever deflection fixed is plotted as a function of (x, y) to form an image. AFM can measure structures with atomic resolution, and new, high-aspect-ratio tips permit imaging of deeply channeled structures (Keller and Chih-Chung, 1992). Examples of the imaging capabilities of the AFM range from atomic resolution imaging of graphite (Fig. 6) (Rugar and Hansma, 1990) to obtaining larger scale views of step structures on an annealed Si surfaces (Fig. 7) (Suzuki et al., 1992) to the visualization of organic layers (Fig. 8) (Garnaes et al., 1992).

The highly oriented pyrolytic graphite (HOPG) surface as well as other layered materials have seen widespread use in STM and AFM studies for instrument calibration and as thin film substrates, since clean, flat, inert surfaces can be prepared by simply cleaving in air. The image of the HOPG surface shown in Fig. 6 demonstrates that the AFM can provide the same high lateral resolution as STM. Furthermore, it should be noted that the symmetry of a graphite image taken with an STM differs from that taken with an AFM because the STM probes the electronic structure, while the AFM "feels" the repulsive interaction between tip and surface.

On a larger scale, the AFM has been used to determine the organization of steps on annealed vicinal Si surfaces. DC heating of Si surfaces flashed to 1200 °C followed by annealing around the $(1\times1)\Rightarrow(7\times7)$ transition temperature ($T_c \approx 885$ °C) had previously resulted in the bunching of steps. The microstructure of these regions obtained by AFM in air, after annealing above or below T_c in ultrahigh vacuum, is shown in Fig. 7 (Suzuki et al., 1992). Step-band regions, interspersed by ter-

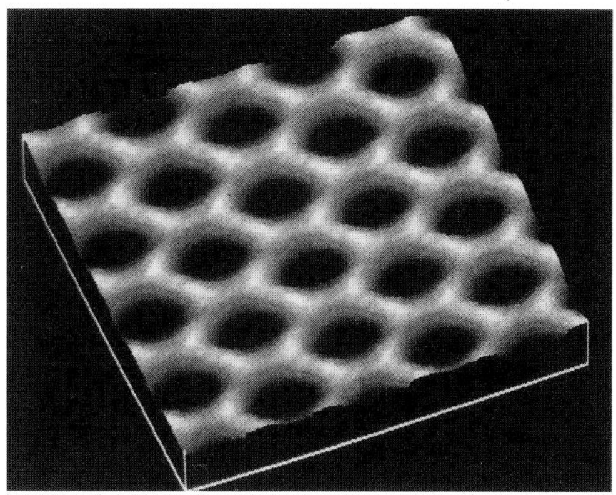

Figure 6. AFM image of a graphite surface; the C atoms are separated by 1.5 Å (Rugar and Hansma, 1990).

Figure 7. AFM images of vicinal Si(111) surfaces flashed to 1200 °C and annealed at 895 °C for 5 min (a), or at 875 °C for 5 min (b), showing step-bunching structures (Suzuki et al., 1992).

races, are of the order of 1 μm in width. Annealing below T_c results in the formation of a large number of sub-step-bands, each several monatomic steps high. These make up the step-band, and the density of sub-steps is higher in the middle than on the sides. At the edges, a curvature of the terraces is observed. Annealing above T_c gives a small number of discontinuous step structures with sharp transitions to the terraces. In both cases, the terrace structures are similar. The curvature obtained at lower temperatures arises due to the formation of sub-step-bands, and is associated with the fact that the $(1 \times 1) \Rightarrow (7 \times 7)$ transition occurs at lower temperatures for larger misorientation angles.

Imaging surfaces of polymer films with AFM provides information on polymer branching, conformation, and intermolecular interactions, without the requirement for electrical conductivity associated with STM imaging. Due to several possible technological applications of Langmuir–Blodgett films at the sub-micron to nanoscale regimes, their molecular structure and packing is a topic of central interest. Figure 8 shows the structure of a four-layer film of cadmium arachidate in which terminal methyl groups exhibiting a height corrugation of < 0.2 nm are imaged (Garnaes et al., 1992). The observed boundary between two distinct ordered domains is found to be misoriented by $\sim 60°$ and the lattice structure is preserved almost to the edge of the boundary. The fact that such ordering is observed leads to the conclusion that the misorientation is due to hexagonal twinning. In addition, the AFM could detect a larger scale buckling in these films which appears to be an equilibrium superstructure.

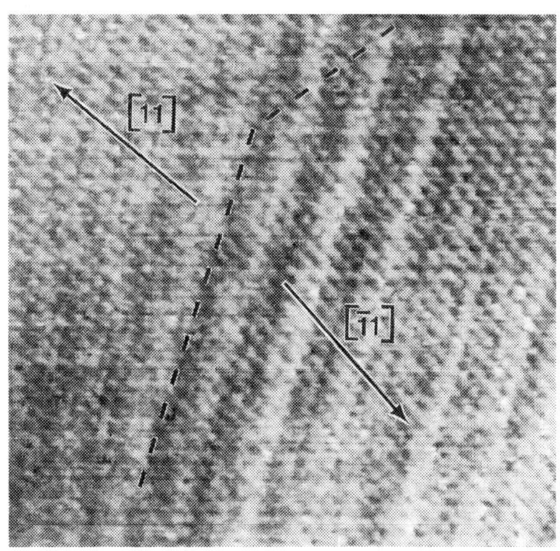

Figure 8. 17.5×17.5 nm^2 AFM images of cadmium arachidate Langmuir–Blodgett films. The dashed line shows the boundary between two domains misoriented by $\sim 60°$ (Garnaes et al., 1992).

Here the techniques of STM and AFM are reviewed by discussing applications in surface science and extensions to other scientific and technological areas in the first two sections. For both microscopies, we begin with a brief historical perspective and cover the theoretical and instrumental issues. The discussion is followed by a section that reviews some of the uses of these techniques for nanoscale manipulation, followed by a section that describes complementary scanning probe techniques based on the design and measurement philosophies of STM.

2 Scanning Tunneling Microscopy (STM)

2.1 Historical Perspective

The major set of advances that led to the invention of the STM was the solution of three long-standing experimental problems. First, although theory could explain the physics of tunneling across a vacuum gap, obtaining the stability necessary to maintain a gap of the order of a few angstroms had not yet been demonstrated. Previously, tunneling measurements were performed using thin, rigid insulating junctions between metal electrodes. Second, the technology to position and scan a probe over a surface with sub-angstrom precision was available using piezoelectric transducers but needed to be refined for this application. Third, a methodology had to be created to bring a sample starting far away from the tip to within ~5 Å without damaging the tip or the sample. This was first accomplished by using a piezoelectric/electrostatic "louse" for coarse approach and the z-piezoelectric transducers for fine z-motion (Binnig et al., 1982 a).

The invention of the STM was preceded by experiments to develop a surface imaging technique whereby a non-contacting tip would scan a surface under feedback control of a field emission current between tip and sample. The manifestation of this concept led to the invention of the *Topografiner* developed at the National Bureau of Standards (NBS, presently National Institute of Standards and Technology) in the late 1960s (Young et al., 1972). The final Topografiner design was remarkably similar to that of the STM ultimately invented by Binnig and Rohrer over 10 years later (Binnig et al., 1982 a, b). At NBS the goal was to develop an instrument for sub-micron scale metrology of surfaces – essentially a non-contacting, high-precision stylus profilimeter. The Topografiner (Fig. 9 a) used a sharp, mechanically-etched tungsten tip which was scanned about 100 Å above the surface with a sufficiently high bias to operate in the field-emission regime with electron current flowing from tip to sample. Feedback control of the vertical piezoelectric actuators was implemented to maintain constant emission current, and the surface topography was obtained by using the z-piezo bias (z-drive voltage) for height as a function of the ramped x- and y-drive voltages. The concept was demonstrated as workable, as illustrated by Fig. 9 b, where the topographic profile of a diffraction grating was mapped. However, the Topografiner was plagued by stability problems due to the difficulty of suppressing vibrations causing instabilities in the tip–sample separation. In their original paper (Young et al., 1972), however, the authors envisioned the eventual progression to imaging in either the field-emission or vacuum tunneling regimes. Curves of current versus separation from the field emission to point contact regimes were measured at various biases. Around the same time, parallel experiments performed at IBM (Thompson and Hanrahan, 1976) using thermal drives and lead whiskers were directed at similar goals.

In a 1982 paper appearing in *Applied Physics Letters,* Binnig and Rohrer reported the first observation of electron tunneling across a stable vacuum gap for several different tip–sample biases (Binnig et al., 1982 b). The vibration problems had been solved by mounting the STM on a magnetically-levitated platform, as seen in Fig. 10, to decouple high-frequencies with the vacuum system resting on a massive bench supported by air cushions. A tungsten tip mounted on a piezoelectric transducer was used to approach a platinum surface, and the exponential dependence of current versus gap separation over four orders of magnitude was observed. This data, reproduced in Fig. 11, confirmed that vacuum tunneling had been achieved. Using the Fowler-Nordheim expression for tunneling resistance, R, as a function of separation, s,

$$R(s) \approx \exp(A\,\phi^{1/2}\,s) \qquad (1)$$

$A = 1.025\ \mathrm{eV}^{-1/2}\,\mathrm{Å}^{-1}$, work function values $\phi \approx 3.5$ eV were obtained for moderately

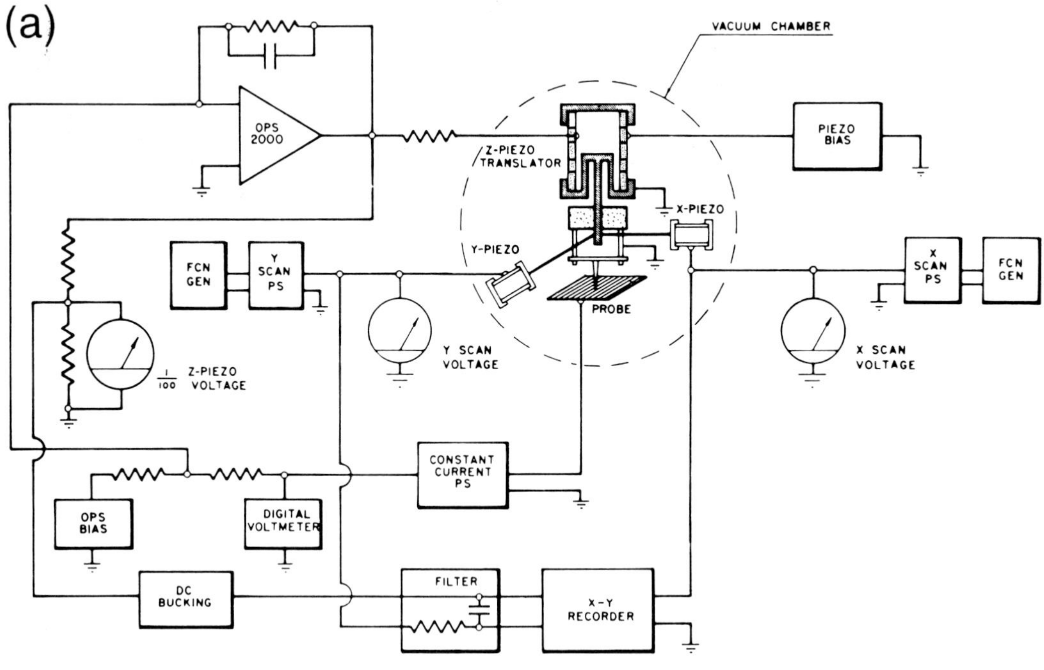

Figure 9. (a) Schematic diagram of the Topografiner; (b) topographic view of a 180 line/mm diffraction grating replica obtained in the FIM mode (Young et al., 1972).

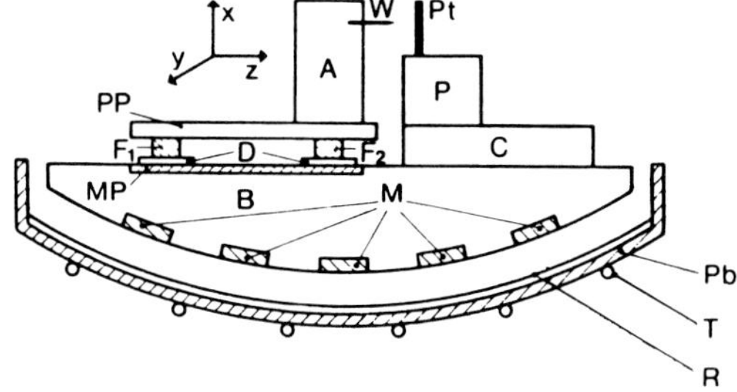

Figure 10. First tunneling unit mounted on magnetic levitation system. The tungsten tip (W) was fixed on a support (A) set on a piezoplate (PP). The electrostatic clamping mechanism consists of feet (F), dielectric sheets (D) and the metal plate (MP). Magnets (M) mounted on the base of the tunneling unit are repelled by the liquid-helium-cooled lead bowl (Pb) (Binnig et al., 1982 a).

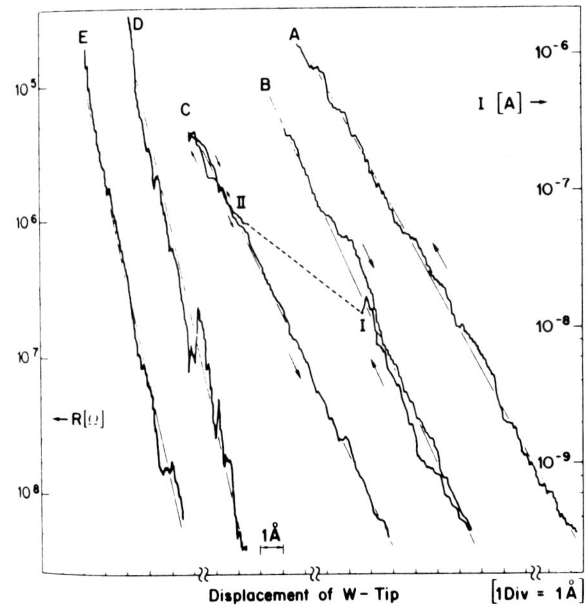

Figure 11. Tunneling resistance (left axis) and tunneling current (right axis) versus vertical displacement of a tip providing first evidence of vacuum tunneling (Binnig et al., 1982b).

clean Pt surfaces. Since metal–insulator–metal tunneling typically gave barriers of <1 eV, the measured value of ϕ confirmed vacuum tunneling. All the essential elements were in place to scan and image, and, in their closing statements, Binnig and Rohrer stated their goal to apply vacuum tunneling as a method for spatially-resolved imaging and electron spectroscopy.

Shortly thereafter, Binnig and Rohrer demonstrated the STM as an atomic-resolution microscope with images of monoatomic steps and reconstructions of the $CaIrSn_4$(110) and Au(110) (Binnig et al., 1982a) surfaces, and the first real-space images of the Si(111)-7 × 7 surface (Binnig et al., 1983b), as shown in Fig. 12. Imaging was accomplished by scanning the tip over the samples while maintaining a constant tunneling current using feedback to control the tip–sample distance. Coarse approach was accomplished by means of a "louse", or piezoelectric-electrostatic walker (Fig. 3). A piezoplate rests with three metal feet insulated from a metal base by a high-dielectric constant material. The feet were clamped on the plate by a high electric field so that elongation and contraction of the piezoplate with an appropriate foot clamping sequence provided for movement in two dimensions.

The image of the Si(111)-7 × 7 surface shown in Fig. 12a is a plot of the z-piezo voltage as a function of x and y positions obtained with a chart recorder. The gray scale image depicts 12 Si adatoms in each unit cell, which are the basis for the reconstruction. The impact of such a capability was immediately evident. For example, the structure of the Si(111)-7 × 7 surface had been a puzzle for over 20 years. Complemented by TEM studies (Takayanagi et al., 1985), this work and subsequent studies provided key information to finally arrive at the actual structure, which is discussed in greater detail in a later section.

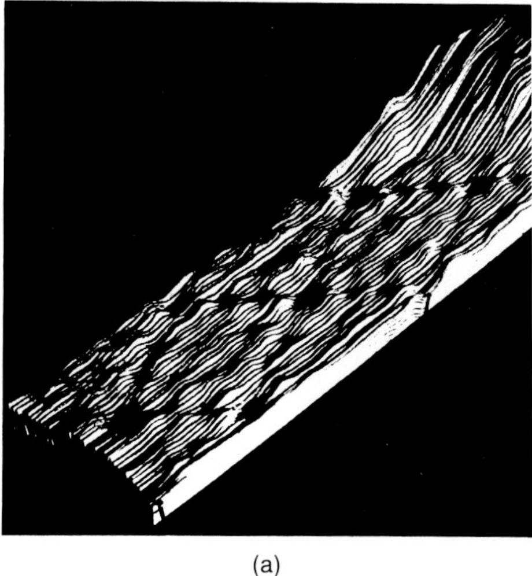
(a)

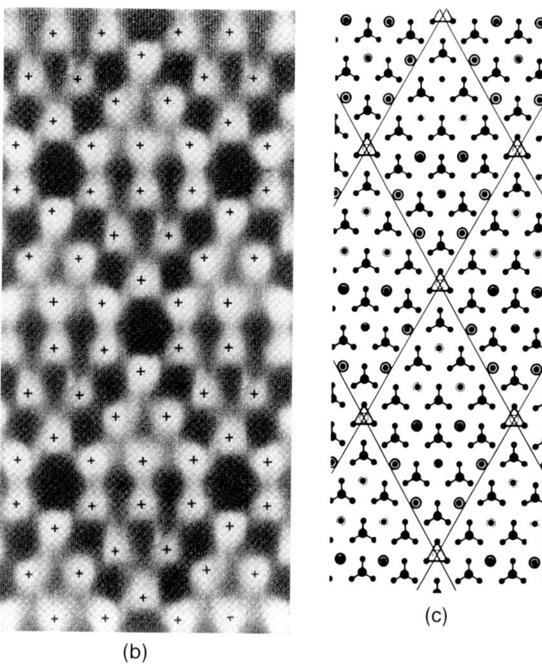
(b) (c)

Figure 12. First topographic image of the Si(111)-7 × 7 surface presented as (a) a topographic relief map and (b) a top-view gray-scale map. (c) The structural model based on these observations (Binnig et al., 1983 b).

2.2 Theory

The most striking aspect of an STM image is that features separated by distances of the order of an interatomic spacing can be resolved. Such features are often taken to represent the position of atoms, and, to the extent that this is true, the STM may be viewed as a tool for measuring the atomic structure of surfaces. However, taking this view without paying attention to the electronic structure has proven to be dangerous for a number of structures. Upon consideration of the tunneling process and the nature of electronic states at the surface and the tip however, it is clear that an STM image really represents spatial variations of the local electronic structure and tunneling barrier at the surface. Our previous comparison of two images of the GaAs(110) surface obtained with biases of opposite polarities (Fig. 4) clearly demonstrated the importance of spectroscopic effects in image interpretation. From this fact it follows that atomically-resolved electronic (and vibrational) spectra can be obtained by acquiring $I-V$ profiles at various locations within a surface unit cell of clean well-ordered regions or in the vicinity of defects or adsorbates, so that a map of electronic states at the surface can be derived.

Here, tunneling between two conductors in one dimension is considered and extended to three dimensions to reflect the geometry of a spherical tip above a corrugated conducting surface. We also consider the transition from the tunneling to the point contact regime, and operation at high biases in field-emission for which electron interferometry is observed. The discussion will address details of the electronic structure of the sample and the tip, so as to understand spatial imaging of localized electronic states at surfaces, including considerations of local inelastic tunneling and ballistic electron emission microscopy (BEEM), where spatial and spectroscopic details of buried interfaces can be obtained.

2.2.1 Electron Tunneling and STM Imaging

Depending on the imaging mode at a given tip bias, an STM image is a map of the z-height of the tip above the surface needed to maintain a constant tunneling current (constant current mode), or of the tunneling current while the tip is scanned at a constant height above the surface (constant height mode). Theoretical models of the tunneling probability provide an understanding of the observed spatial resolution and spectroscopic aspects of STM. The first theoretical discussion of tunneling from a point electrode was presented by Tersoff and Hamann (1983), which was based on the earlier work of Bardeen (1961). Development of the one-dimensional formalism begins with the calculation of the tunneling current between the wavefunctions (WFs) of a metal tip (T) and a metal surface (S), as depicted in Fig. 13. The geometry for the calculation is shown in Fig. 14. In first-order perturbation theory,

$$I = \frac{2\pi e}{\hbar} \sum_{TS} f(E_T)\,[1 - f(E_S + eV)]\,|M_{TS}|^2 \cdot \delta(E_T - E_S) \quad (2a)$$

where

$$f(E) = \exp\left\{\frac{1}{\left[\frac{(E-\mu)}{(k_B T)} + 1\right]}\right\} \quad (2b)$$

i.e., $f(E)$ is the Fermi function, V is the applied voltage, M_{TS} the tunneling matrix element between the tip WF (ψ_T) and the surface WF (ψ_S), $\delta(x)$ the delta function, E_i the energy of a state in the absence of tunneling ($i = $ T, S), and μ the chemical potential.

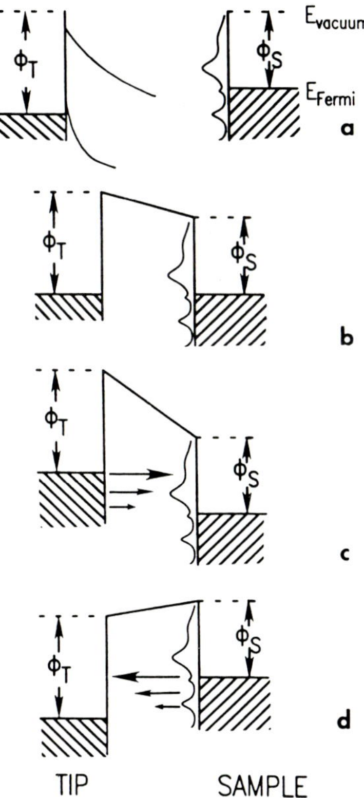

Figure 13. Energy level diagrams of the sample and the tip: (a) with the sample and tip independent, the vacuum levels are equal; (b) with the sample and tip connected with no potential difference, the Fermi levels are equal; (c) application of a negative tip bias, $V_t < 0$, allows electron to flow from occupied states of the tip to unoccupied states at the sample surface; (d) application of a negative sample bias, $V_t > 0$, allows electrons to flow from occupied states of the sample surface to unoccupied states in the tip (Hamers 1989).

For small tunneling voltages (e.g. ~10 meV between the tip and the substrate) and low temperatures ($k_B T$ ~ 50 meV for $T = 300$ K), $f(E)$ can be expanded to obtain

$$I = \frac{2\pi}{\hbar} e^2 V \sum_{TS} |M_{TS}|^2\, \delta(E_T - E_F) \cdot \delta(E_S - E_F) \quad (3)$$

The matrix element M_{TS} is obtained from

$$M_{TS} = -\frac{\hbar^2}{2m}\int \mathrm{d}\boldsymbol{S}\cdot(\psi_T^* \nabla\psi_S - \psi_S \nabla\psi_T^*) \tag{4}$$

where m is the free electron mass, the integral is taken over a surface lying entirely within the vacuum barrier region between the tip and the sample, * denotes the complex conjugate, and

$$\nabla = \mathbf{i}\frac{\partial}{\partial x} + \mathbf{j}\frac{\partial}{\partial y} + \mathbf{k}\frac{\partial}{\partial z}$$

Carrying out the integral in one dimension (plane waves) by taking ψ_T^0 and ψ_S^0 as the wavefunctions of the tip and the sample, respectively, gives the explicit exponential dependence of the tunneling current as a function of tip–sample separation. In particular, if the tip and sample are identical materials, the wavefunctions in the vacuum gap can be written as

$$\psi_S(z) = \psi_S^0\,\mathrm{e}^{-\varkappa z} \tag{5a}$$

$$\psi_T(z) = \psi_T^0\,\mathrm{e}^{-\varkappa(d-z)} \tag{5b}$$

where $\varkappa$ is the inverse decay length, i.e.,

$$\varkappa = \frac{2m\phi^{1/2}}{\hbar^2} \approx 0.512\,\phi^{1/2} \tag{6}$$

with ϕ, the average work function, in electron volts and $\varkappa$ in Å^{-1}. Then the tunneling current exhibits an exponential dependence on the separation d given by

$$I \propto \sum |\psi_S^0|^2\,|\psi_T^0|^2\,\mathrm{e}^{-2\varkappa d} \tag{7}$$

Continuing with the spherical tip configuration of Fig. 14, M_{TS} may be evaluated to assess the potential spatial resolution of STM. The surface wavefunctions are expanded in plane waves parallel to the surface with decaying amplitude into the vacuum

$$\psi_S^0 = \Omega_S^{-1/2}\sum_G a_G \exp[-(\varkappa^2 + |\boldsymbol{k}_{\|} + \boldsymbol{G}|^2)^{1/2} z] \cdot \exp[i(\boldsymbol{k}_{\|} + \boldsymbol{G})\cdot \boldsymbol{x}] \tag{8}$$

where Ω_S is the sample volume, $\varkappa = (2m\phi)^{1/2}/\hbar^2$, $\boldsymbol{k}_{\|}$ is the surface wavevector, and $\boldsymbol{G}$ is a two-dimensional reciprocal lattice vector of the surface.

(a)

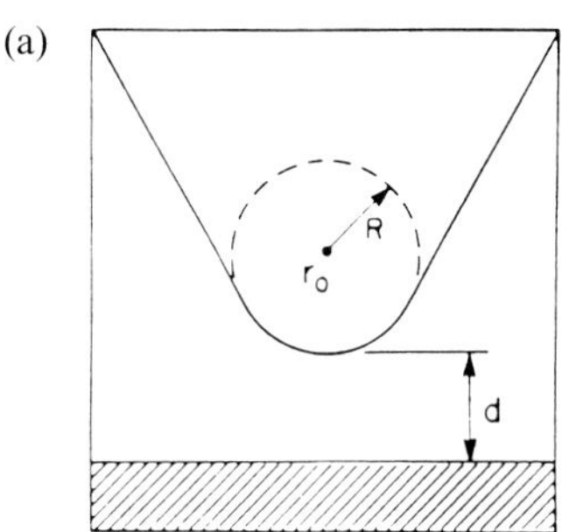

(b)

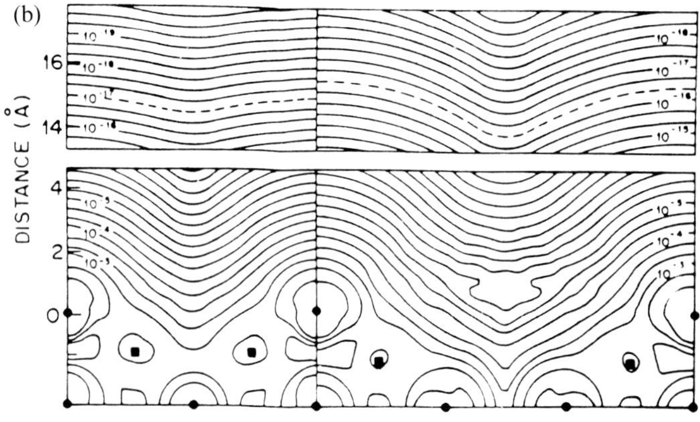

Figure 14. (a) The geometry for tunneling microscopy calculations shows the tip radius, R, and the tip apex-to-sample distance, d. (b) Contours of constant charge density at the Fermi level, $\varrho(r, E_F)$, are presented for the Au(100) (2 × 1) (left) and Au(100) (3 × 1) surfaces (Tersoff and Hamann, 1983).

The spherical tip ($R \gg k^{-1}$) wavefunctions are expanded in similar form

$$\psi_T^0 = \Omega_T^{-1/2}\, c_T\, \varkappa R\, e^{\varkappa R} (\varkappa |\boldsymbol{r} - \boldsymbol{r}_0|)^{-1} \cdot \\ \cdot e^{-\varkappa |\boldsymbol{r} - \boldsymbol{r}_0|} \tag{9}$$

where Ω_T is the probe volume and R is the radius of curvature of the tip. The resulting matrix element is

$$M_{TS} = -\frac{\hbar^2}{2m} 4\pi \frac{(\varkappa R\, e^{\varkappa R})}{(\varkappa \Omega^{1/2})} \psi_S(\boldsymbol{r}_0) \tag{10}$$

where $\boldsymbol{r}_0$ is the position of the center of curvature of the tip. The tunneling current is

$$I = \frac{32\pi^3 e^2 V}{\hbar} \phi^2 D_T(E_F) R^2 \varkappa^{-4} \cdot \\ \cdot e^{2\varkappa r} \sum_S |\psi_S(\boldsymbol{r}_0)|^2 \delta(E_S - E_F) \tag{11}$$

where ϕ is the work function of sample and tip, and $D_T(E_F)$ is the tip density of states per unit volume.

Without using exact wavefunctions for the tip and the surface, the spatial resolution of the STM was approximated by examining the inverse decay length demonstrating a lateral resolution for Au(110)-2×1 of ≤ 5 Å (Tersoff and Hamann, 1983), in agreement with experiment (Binnig et al., 1983a). Another theoretical investigation (Stoll, 1984) of the relation between resolution, tip diameter ($2r$), and tip–sample distance (s) produced a relationship for vertical attenuation versus lateral period, which is depicted in Fig. 15. The vertical attenuation refers to the ratio of the measured corrugation versus the real corrugation of a metal surface for various values of $r+s$. This figure shows quantitatively that, if $r+s$ increases, the instrument has a reduced sensitivity to the surface corrugation, as expected.

In addition, for small bias voltages, the tunneling conductance, $\sigma = I/V$, can be re-

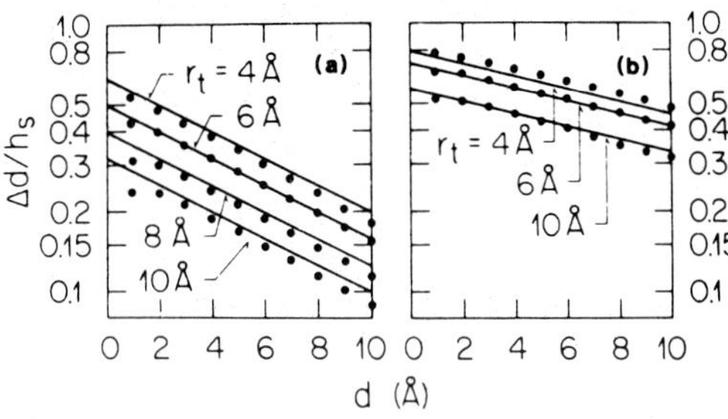

Figure 15. A set of curves depicting the sensitivity to corrugation amplitudes as the tip radius (r_t) or tip–sample separation are increased after a theoretical treatment (Stoll, 1984). This is given by the vertical attenuation plotted as a function of the lateral period.

lated to the density of states ϱ at the surface for energies around E_F for a particular tip position. That is,

$$\sigma \propto \varrho(\boldsymbol{r}_0, E) \tag{12a}$$

where

$$\varrho(\boldsymbol{r}_0, E) = \sum_S |\psi_S(\boldsymbol{r}_o)|^2 \delta(E_S - E_F) \tag{12b}$$

This observation will become important when spectroscopic capabilities of the STM are discussed below.

The Local Barrier Height

The exponential dependence of the tunneling current depends on both the tip–sample separation and the local barrier height. Typically, it is assumed that the barrier height is constant over the entire scanned area, but, if an adsorbed patch of impurities changes the work function in a particular region on the sample, the tip will respond to the reduced barrier height and a structure unrelated to the topography will appear in the image. Recognizing that $A\,\phi = \mathrm{d}(\ln I)/\mathrm{d}s$, the position dependence of the barrier height can be extracted independently during a scan. In the constant current mode, this can be accomplished by modulating the tip distance by Δs at a frequency higher than the cutoff frequency of the feedback circuit while scanning (Binnig and Rohrer, 1983; Wiesendanger et al., 1987). In this way, the work function analysis and feedback image acquisition are separated. Using lock-in detection of the current signal, $d(\ln I)/ds$ provides the desired position-dependent information on the work function in parallel, with the constant current topographic image.

The Transition from Tunneling to Point Contact Regimes

Besides tunneling, two conduction regimes exist for close tip–surface distances – electronic contact and point contact. Although these regimes can be modeled theoretically (Lang, 1987b; Ciraci and Tekman, 1989), they are difficult to control in practice because the exact shape of the tip and the degree of the tip interaction with the surface by, e.g., van der Waals or other forces are not well-defined. Defining the separation as between atomic centers, structural instabilities can be assumed to occur when the separation is equal to the sum of the atomic radii. The *electronic contact* regime refers to distances for which the tip is sufficiently close to the sample to affect the respective wavefunctions; this leads to the appearance of tip-induced energy sub-bands above E_F (Binnig et al., 1985). *Point contact* refers to the collapse of the potential energy barrier with the onset of ballistic transport due to the creation of energy sub-bands below E_F. It has been shown experimentally that tunneling at a reduced barrier height can be maintained for close approach until an oscillatory $I-V$ characteristic appears beyond some structural instability (Gimzewski et al., 1987). The $I-z$ characteristic for an iridium tip and a polycrystalline Ag surface, shown in Fig. 16a, exhibits an abrupt increase in tunnel current interpreted as the onset of

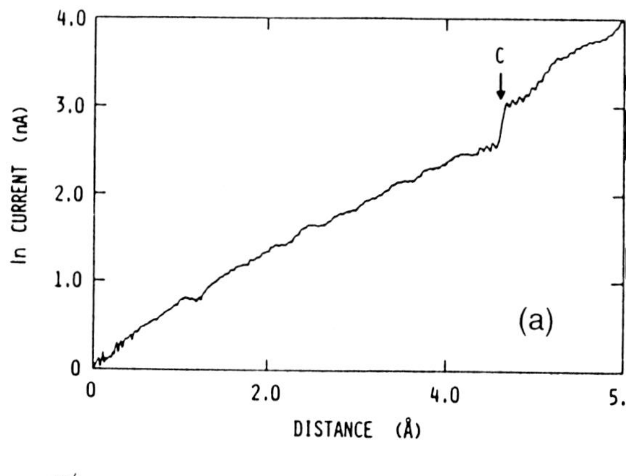

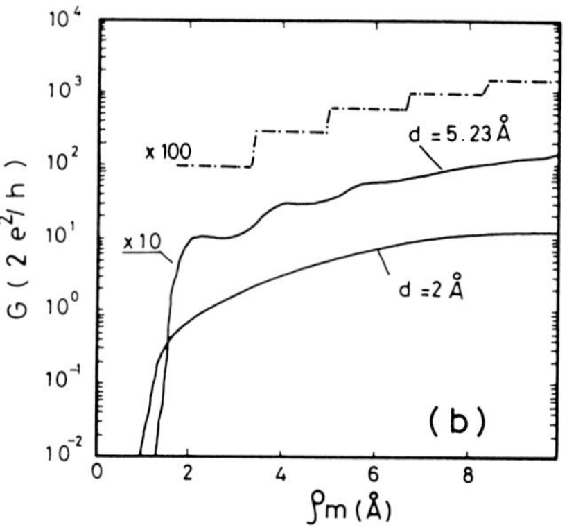

Figure 16. (a) Tunneling current versus tip–sample separation for a clean Ir tip and an Ag surface (Gimzewski et al., 1987). (b) Conductance, G, versus the diameter, ϱ_m, of the orifice set up by a tip atom for various values of separation, where $\lambda_F = 5.23$. Quantization steps are predicted at distances of the order of λ_F, the Fermi wavelength (Ciraci and Tekman, 1989).

electrical contact. Analysis of this abrupt change in resistance indicates the contact area of a single atom. Contact can lead to reversible or irreversible local surface modifications. Calculations (Ciraci and Tekman, 1989) for the close approach of an Ag tip (single atom at the vertex) towards an Ag surface show the details of such a transition (Fig. 16b). At a separation of the sum of the atomic radii, the conductance saturates around $2e^2/h$, representative of the ballistic transport limit. As the radius of contact increases, steps in the conductance appear. Beyond the point contact regime, the system settles into a more characteristic conductance versus separation curve. However, in the electronic regime, the barrier height is still varying, whereby the first tunneling sub-band dominates the conductance.

Density of States Effects

The effect of electrode densities of states on tunneling images and their spectroscopic signatures was explored by the consideration of the spatial dependence of the tunneling current between a single atom and a planar surface (Lang, 1985) and variations thereof (Lang, 1986, 1987a). The electrodes consisted of an alkali atom adsorbed on a planar jellium surface and a second planar jellium surface (Lang, 1985). For Na adsorbed on jellium it had been established that the Na 3s-level crosses E_F and is partially filled due to charge donation from the substrate (Lang and Kohn, 1970). By modeling the details of tunneling for this system, two key results were obtained, as shown in Figs. 17 and 18. First, tunneling current density contours show that electron flow is localized about the adsorbed atom and is significantly enhanced by the protrusion of the atom from the substrate. Second, tunneling current increases with increasing state density in the vicinity of E_F. The state densities of an Na adatom and a Ca adatom on jellium are shown in Fig. 17. Figure 18 shows a contour map of j_z/j_0 (where j_z is the current density along z and j_0 is the current density in the absence of an adsorbed atom) for the two substrates separated by 16 Å with the adsorbed Na atom at its computed equilibrium distance. The current density is localized about the adsorbed atom as a result of the state density of the adatoms around E_F. Due to the spatial nature of various states, $m = 0$ (s and p_z) states provide the major

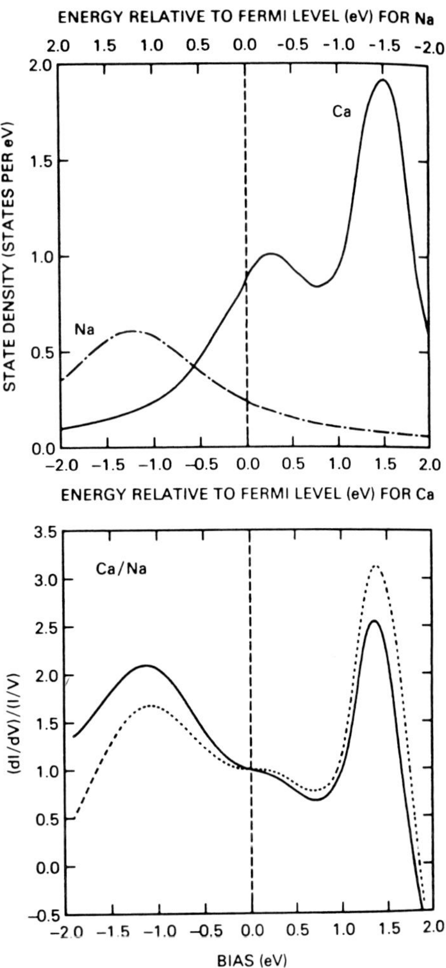

Figure 17. Upper panel: the difference in eigenstate density between a metal–adatom system and the bare metal for Ca and Na. The Na 3s and Ca 4s and 3d/4p contributions are observed. Lower panel: calculated curves of $(dI/dV)/(I/V)$ versus V for the Ca/Na system (solid), and with the tip and sample state densities constant (dotted) (Lang, 1986).

contribution to the tunneling current. Ca, as compared to Na, has a greater state density near E_F, and results in an increase in current directly above the adatom by a factor of ~ 4 (Lang, 1985).

For tunneling at a bias representative of a specific electronic state, the current forming the image will be dominated by that state (and those of lower biases). The effect

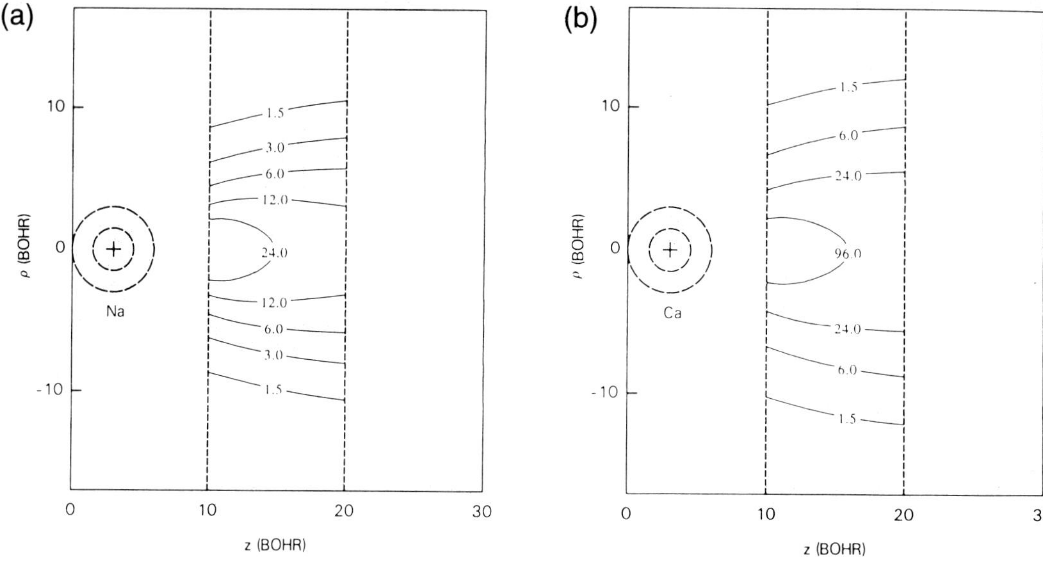

Figure 18. (a) Contour map of j_z/j_0 for a Na atom protruding from the left electrode. (b) Contour map of j_z/j_0 for a Ca atom protruding from the left electrode (Lang, 1985).

of different states on the vertical displacement has been demonstrated by calculations for Na, S, and Mo (Lang, 1987a). It was found that the dominant contributions are for s- and p-states, but that d-states give almost no contribution. In addition, the origin of negative displacements of the tip for regions where there is reduced state density atop an adsorbate were also discussed.

Having shown the localized nature of the tunneling current and making an analogy to tunneling spectroscopy of metal–insulator–metal structures, the derivative of the tunneling current as a function of tip–sample bias (dI/dV) can be expected to give a function related to the joint density of states (JDOS) localized within atomic dimensions. Figure 19 shows the result of such a calculation in which $(dI/dV)/(I/V)$ is compared for tunneling between a tip and a sample of constant state density and a Ca sample with a constant state density tip. The error as a function of the width of the electronic state of the sample being probed is found to be smaller with $(dI/dV)/(I/V)$ than with (dI/dV). $(dI/dV)/(I/V)$ is routinely plotted for spectra obtained with the STM.

2.2.2 Scanning Tunneling Spectroscopy (STS)

Occupied and Unoccupied States near E_F

The previous discussion initially considered situations for which the charge density follows the atomic corrugation at the surface. For example, the contour of s-orbitals would reflect the atomic positions of metal surface atoms, but the extended nature of the s-band in the surface plane would result in extremely small corrugation amplitudes. Semiconductors, on the other hand, possess directional covalent bonds; at the surface, these occupied or unoccupied electronic states may extend out into the vacuum and may reside energetically in the bulk and gap. Assuming a

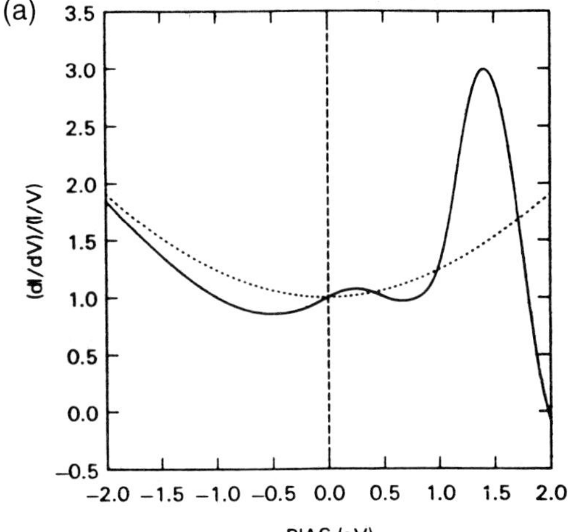

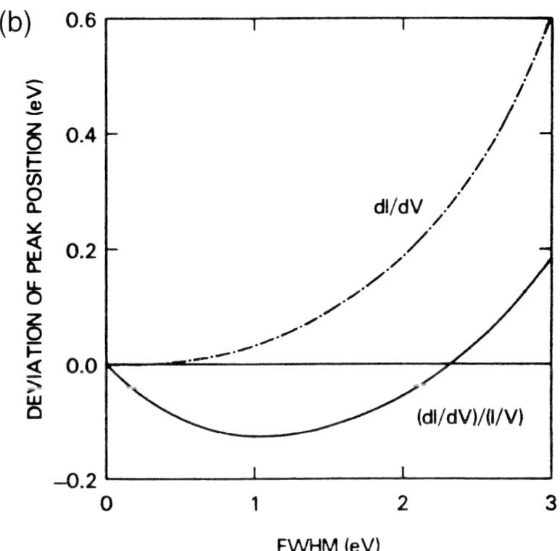

Figure 19. (a) Comparison of the calculated density of states as given by the function $(\mathrm{d}I/\mathrm{d}V)/(I/V)$ for a tip of constant state density and (dotted) a surface with constant state density and (solid) a surface describing the density of states of Ca. (b) The deviations of peak positions of the $\mathrm{d}I/\mathrm{d}V$ (dashed) and $(\mathrm{d}I/\mathrm{d}V)/(I/V)$ (solid) with respect to the actual maxima of the density of states (Lang, 1986).

simple metal tip, the tunneling current is derived from states at the sample surface between E_F and $(E + e\,V_t)$. With the goal of mapping localized electronic states related to possible reconstructed phases or defects, an appropriate tip–sample bias must be applied so as to "tune into" specific state(s). For simplicity we assume that the metal tip has a fairly constant density of states (DOS). The previously discussed comparison of constant-current images of the GaAs(110) surface with opposite polarities described the spatial localization of the occupied and unoccupied states at the surface (Feenstra et al., 1987b). As a next step, recording I–V relations with the STM by sweeping the bias at each atomic site defines the principle of scanning tunneling spectroscopy (STS). STS allows the measurement of the energy levels at localized positions on the surface, which can be processed to map locations of individual surface electronic states.

The use of STM as a local spectroscopic probe (Hamers, 1989; Tromp, 1989) has expanded particularly for the study of semiconductor surfaces and interfaces. Theoretically, the JDOS of tip and sample can be taken into account by explicitly including their functional dependencies in the tunneling matrix element, as previously described (Lang, 1986). In an early treatment of bias-dependent tunneling, the spatial characteristics of σ- and π-bands of graphite as a function of tunneling bias were calculated (Selloni et al., 1985). The calculated corrugation amplitudes of individual wavefunctions in the vacuum region demonstrated that discrete states could be imaged by bias-dependent imaging. It was also shown that band structure effects are important. Layered materials such as graphite possess Fermi surface structures at surface Brillouin zone (SBZ) boundaries for which the surface wavefunctions exhibit standing wave solutions. The corresponding STM images of those states do not then reflect atomic positions, and the corrugation amplitudes do not reflect the atomic corrugation.

Variations in the slope in an I–V profile occur when new states begin to contribute to the tunneling current. The differential conductivity, dI/dV versus V, gives a rough picture of the state density. While there is no simple relation between the intensity of these features and the DOS $(dI/dV)/(I/V)$, or $d(\ln I)/d(\ln V)$, versus V better reflects the positions of the maxima in the DOS (Lang, 1986; Feenstra et al., 1987a). It is advantageous to plot $(dI/dV)/(I/V)$ for other reasons too, i.e., due to the exponential relationship of I and dI/dV versus tip–sample separation, and because dI/dV diverges at zero bias. One means of enhancing sensitivity at low biases is to maintain a junction of constant resistance by reducing the tip–sample distance as the bias is decreased (Feenstra et al., 1987a).

Let us first consider spectra obtained with the STM tip above one point on a surface, as illustrated by the family of $I-V$ curves obtained from an Si(111)-2 × 1 surface for various tip–sample separations in Fig. 20. Several observations should be noted. (1) Zero conductance is observed for a region consistent with the width of the surface band gap, particularly at the minimum tip–sample separation. (2) The $I-V$ curves exhibit rectifying behavior; i.e., the current for positive sample bias is greater than for negative bias. (3) At constant current, the separation versus bias curves exhibit ripples above $\sim \pm 4$ eV. The surface band gap in the Si(111)-2 × 1 surface is derived from vertical transitions between

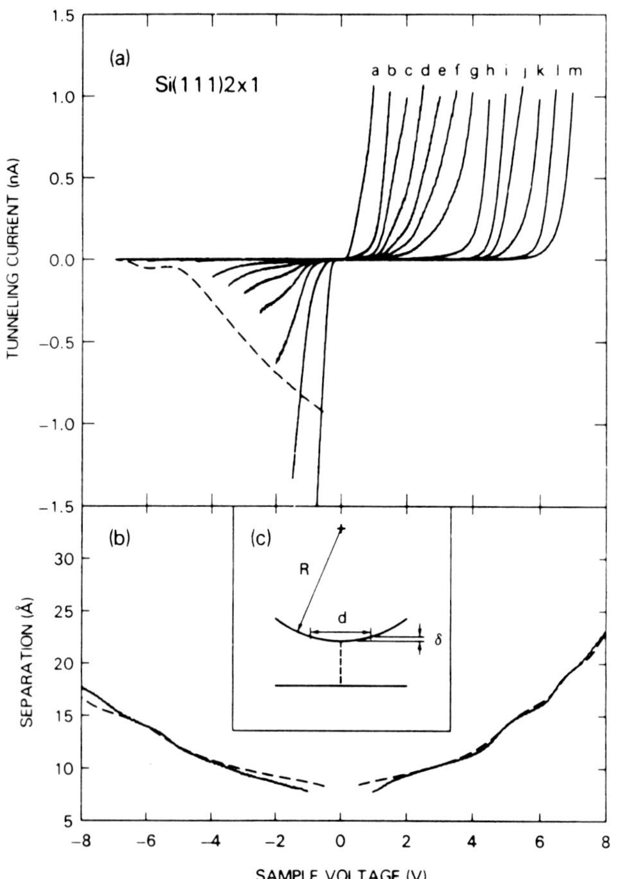

Figure 20. (a) Scanning tunneling spectra of the Si(111)-2 × 1 surface as a function of tunneling distance with the (x, y) position of the tip fixed. Tip sample separations range from 7.8 Å to 19.5 Å for the spectra a through m. (b) The separations were obtained by measurement of separation versus voltage to maintain a 1 nA tunneling current. Theoretical calculations (dashed lines) used the tip geometry given in (c) (Feenstra et al., 1987a).

the π surface bands at the surface Brillouin zone (SBZ). States away from the center of the SBZ decay rapidly into the vacuum; thus, if states at the SBZ center exist, they dominate the tunneling. For Si(111)-2×1, the tip must sample non-zero $q_{\|}$ (electron wavenumber parallel to the surface), and this ability depends on the magnitude of the inverse decay length, k, which is a function of the tip–sample separation and bias voltage. The value of the inverse decay length is >1 Å^{-1} for biases <1 V, so that the corner of the SBZ can indeed be probed. This rectifying behavior can be explained by the effect of the enhancement of the electric field by the tip, which effectively lowers the potential barrier for tunneling into unoccupied states of Si. The barrier resonances occur due to interferometric effects of free electrons at high bias.

When the tip is scanned, a comparison of spectroscopic images over a wide range of sample biases can provide a picture of localized states at the surface. A conductance image was first obtained with an STM on Si(111)-7×7 by modulating the tip–sample bias around the imaging tunneling bias and using lock-in detection to man $\mathrm{d}I/\mathrm{d}V$ (Binnig and Rohrer, 1986). Such images depend on the average bias since this controls the z-height through the feedback loop. Current imaging tunneling spectroscopy (CITS) is an alternative method to obtain spectroscopic images using tip-to-sample reference heights at each position determined by the constant-current imaging bias (Hamers et al., 1986c). This bias is chosen such that it is not overly sensitive to particular surface states. A sample–hold circuit is employed so that, for each pixel, the tunneling current for a series of discrete bias voltages is acquired. Displaying current images at this point only provides indirect information about the surface electronic structure similar to the difficulty in interpreting constant-current images for different biases. However, if *differences* between current images can be displayed, it is possible to display individual electronic states. In fact, if differences between the logarithm of current images for two adjacent biases are plotted, a density of states image midway between the biases is obtained. The choice of the imaging bias still affects the CITS images, and CITS images for a variety of imaging biases typically need to be compared to normalize out this effect.

The first demonstration of CITS provided a dramatic and conclusive local picture of the origin of the surface states representative of the Si(111)-7×7 surface (Hamers et al., 1986c). Previously photoemission (Himpsel and Fauster, 1984) and inverse photoemission (Fauster and Himpsel, 1983; Himpsel and Fauster, 1984) experiments had produced the energy position of these states, but, being techniques that average over the surface, the nature of the electronic states was inferred indirectly. Figure 21 shows three current images obtained for $0.15 < V_t < 0.65$ V, $0.60 < V_t < 1.0$ V, and $1.6 < V_t < 2.0$ V. The three CITS images are different because they each map different surface states – the adatom state, the dangling bond state, and the backbond state, respectively. To further illustrate this assignment, a comparison of CITS spectra taken at different locations in a 7×7 unit cell is also shown in Fig. 21. The "average" of these spectra follows ultraviolet photoelectron spectroscopy (UPS) and inverse photoelectron spectroscopy (IPES) spectra and the position-dependent spectra show the origin of each spectral feature. CITS images, as well as STM images obtained at low biases, show that the faulted and unfaulted halves of the 7×7 unit cell appear different. This shows the contribution of the stacking sequence in defining the elec-

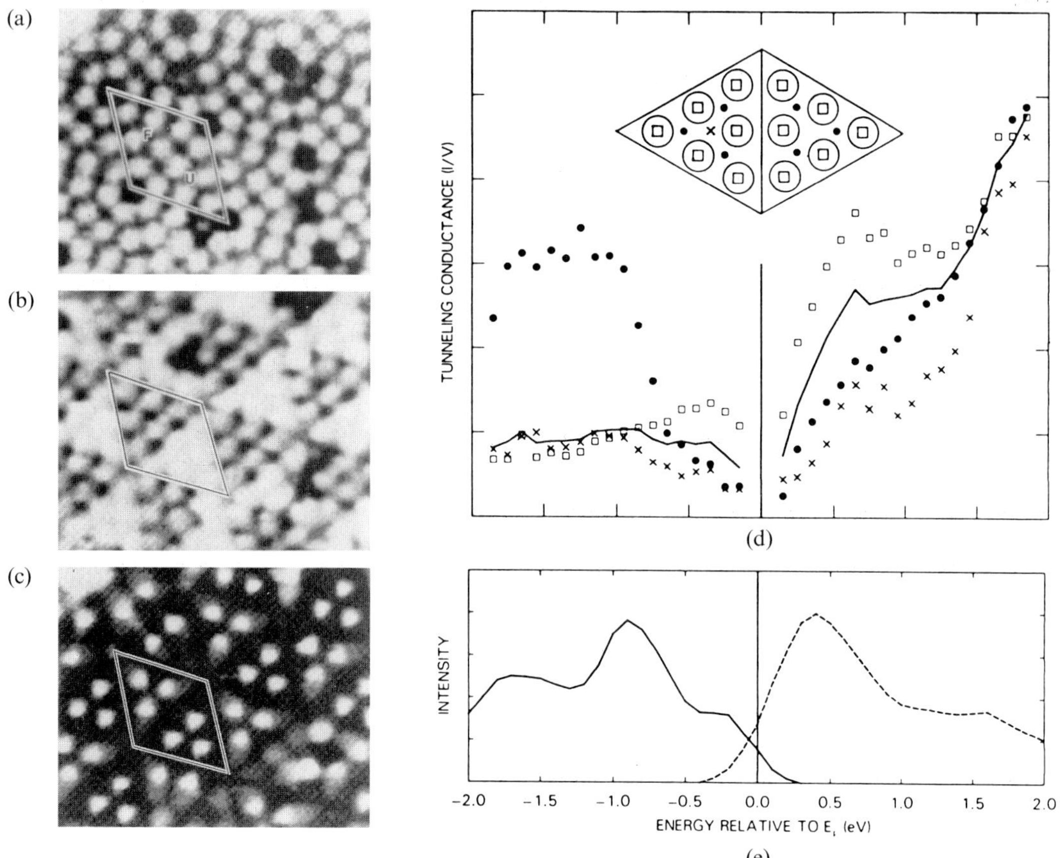

Figure 21. (a)–(c) Local spectroscopic images of the Si(111)-7 × 7 surface obtained in the CITS mode: (a) $0.15\ V < V_t < 0.65\ V$; (b) $0.60\ V < V_t < 1.0\ V$; (c) $1.6\ V < V_t < 2.0\ V$. (d) Position-dependent CITS spectra compared with (e) photoemission (solid curve) and inverse photoemission (dashed curve) spectra (Hamers et al., 1986c).

tronic structure. Similarities in CITS images for both polarities at low bias support the prediction of a metallic state. Observation of the dangling bond states between the adatoms provided further support for the dimer-adatom-stacking (DAS) model (Takayanagi et al., 1985), which contains dangling bonds beneath the adatom layer that are involved in the surface chemistry at the Si(111)-7 × 7 surface.

Measuring electronic spectra with atomic resolution in real space removes the difficulty in separating the contributions of local defects from the intrinsic electronic structure of the ideal ordered surface. For example, Fig. 22 shows a set of spectra taken at adjacent sites near a defect along a dimer row of an Si(111)-2 × 1 surface (Hamers, 1989). The Si(100) surface is discussed in greater detail in the following section. The gap is decreased close to the defect, with the greatest effect at the defect site itself. This example of spatially-resolved spectroscopy immediately reveals that the defect induces electronic states at E_F, resulting in metallic areas separated by semiconducting regions.

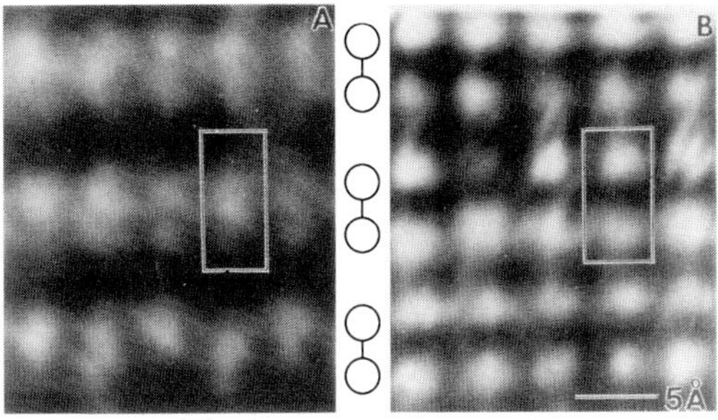

Figure 22. Constant current images of the Si(100)-2×1 surface showing the dimer bond between atoms in the occupied state topographic image ($V_t=1.6$ V), and the localization of charge about the dimer atoms in the unoccupied state current image ($V_t=-1.6$ V) (Hamers, 1989).

Spectroscopy at Higher Biases

Surface electronic states, called image states, are bound states that reflect the response of the substrate in the presence of electrons near the surface. These states are modified by the translational invariance along the surface and the reflective properties of the surface, but form a hydrogen-like series of energies given by

$$E_n = \varepsilon_n + E_{n,\mathrm{corr}} + \frac{\hbar^2 \boldsymbol{k}^2}{2m_n^*} \tag{11-13}$$

where n is the principal quantum number, ε_n is the hydrogenic component of the binding energy, $E_{n,\mathrm{corr}}$ reflects deviations from free electron motion, m_n is the effective mass, and $\boldsymbol{k}$ is the parallel wavevector. One response of the substrate to the biased tip is the creation of an image charge distribution which results in an attractive force between tip and sample (Binnig et al., 1984). In addition, the strong electric field in the gap modifies the image state spectrum by increasing the separation between the levels and extending the image states well above the vacuum level (Binnig et al., 1985). It was shown (Binnig et al., 1985) that these states are observable for high sample biases as strong peaks in the tunneling conductance for metal and semiconductor surfaces. These peaks were shown to follow the expected modified hydrogenic progression extending several eV above the vacuum level. In addition, the peaks shift as a function of field strength (varying the gap), whereby inverse photoemission would correspond to the zero-field limit of the image state spectrum.

For biases in the field-emission regime, where the electron energy is greater than the tunneling barrier height, standing waves can form in the barrier. As a result, the STM acts as an electron interferometer, whereby structure in the tunneling conductance occurs as a function of electron kinetic energy and/or tip–sample separation (Becker et al., 1985a). This is seen in Fig. 23 where the conductance and gap distance between a W tip and an Au surface is plotted versus bias voltage at constant tunneling current to provide a wide dynamic range. Eleven oscillations in $\mathrm{d}I/\mathrm{d}V$ are observed corresponding to standing wave solutions of the electron wavefunctions in the gap, and these were modeled with and without image effects. Beyond the fundamental quantum mechanics of this observation, such measurements are significant because they provide means to obtain the work function and the absolute tip–sample separation. Electron standing-wave interferometry can not only be used between tip and sample but has also been used to form standing waves in ultrathin films to provide contrast at a buried interface. This has been demonstrated for ultrathin $NiSi_2$ films grown on Si(111) (Kubby and Greene, 1992).

2.2.3 Inelastic Tunneling Spectroscopy

The use of STS to probe the occupied and unoccupied electronic states and par-

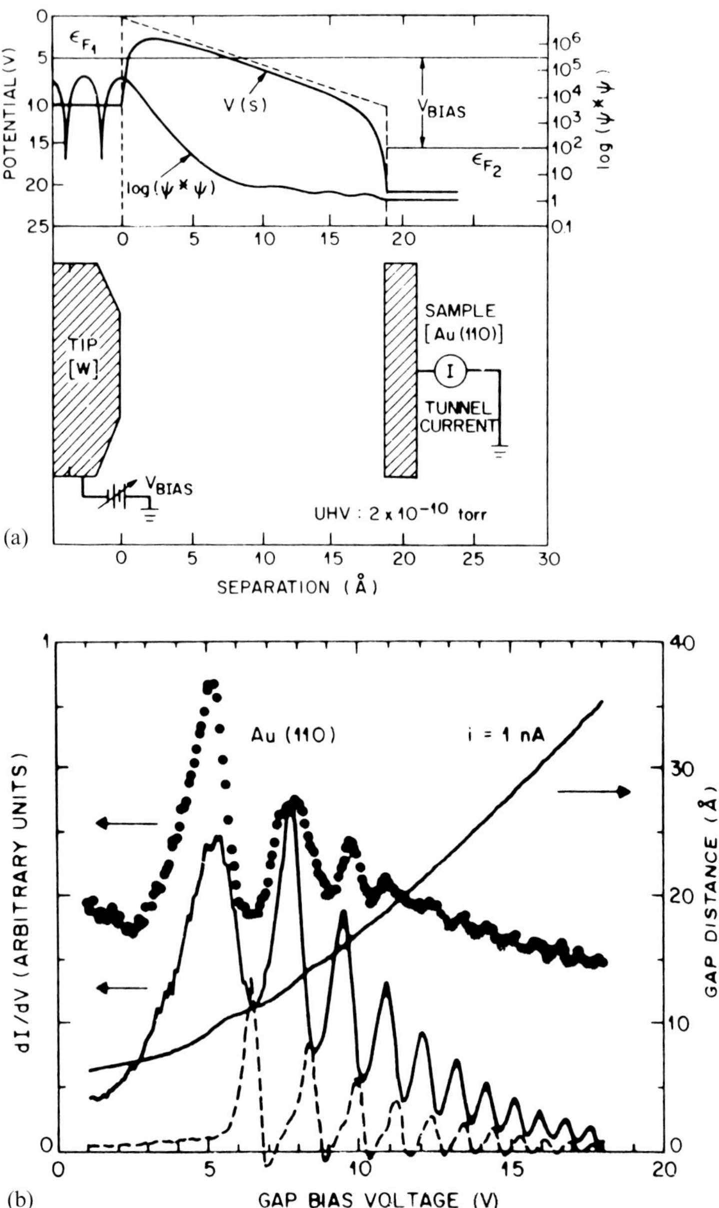

Figure 23. (a) Experimental geometry and potential energy diagram for calculations of standing wave oscillations in the tunneling gap. (b) Measurements of dI/dV and gap distance as a function of bias voltage compared to theoretical curves with (solid) and without (dashed) image potential contribution (Becker et al., 1985a).

ticular sites on a surface involves elastic tunneling through the vacuum gap. Electrons passing through the vacuum tunnel gap may also be expected to produce vibrational or electronic excitations by a process analogous to inelastic electron tunneling spectroscopy (IETS). The localized flow of current from a tip has been proposed as a means to probe individual adsorbates in their local bonding environments. The vibrational interaction has been considered theoretically in this context (Persson and Demuth, 1986; Persson and Baratoff, 1987; Baratoff and Persson, 1988). An inelastic loss would be observed as a change in slope of the differential conductance at an energy referenced to the elastic signal. These types of measurements have the potential to provide the vibrational or electronic signatures of a segment of an adsorbed molecule just beneath the tip, even if it is "invisible" to imaging at a particular bias. In surface chemistry this could provide information on reactions at active sites on a surface which may reside at or near steps or defects.

The theoretical treatment of inelastic tunneling for vibrational excitations can be derived from the formalism similar to that used for electron energy loss spectroscopy (EELS) (Ibach and Mills, 1982), where the inelastic signal intensity is typically $< 1\%$ of the elastic background. Inelastic tunneling demands a high signal-to-noise ratio for spectroscopy. For chemisorbed species on metals, model calculations (Persson and Baratoff, 1987) predicted that adsorbate-induced resonances in the vicinity of E_F could increase the inelastic signal by a factor of ~ 10. In addition, symmetry considerations must be included to determine the coupling of wavefunctions between tip and sample. Furthermore, when the current passing between the tip and the sample is compared to the vibrational relaxation time, it has been noted that many of the adsorbates will be probed in an excited vibrational state. Such demonstrations of inelastic tunneling with an STM have yet to be convincingly demonstrated.

A variation of this concept allows the observation of plasmon modes included by the interaction of electrons emitted from the tip. Inelastic interactions of both tunneling electrons and field-emitted electrons on Ag surfaces have been shown to induce optical emission (Berndt et al., 1991) which, if plotted as a function of the scanned tip position, produces a map of the regions where the inelastic excitations originate.

2.2.4 Ballistic Electron Emission Microscopy (BEEM)

Ballistic electron emission microscopy combines electron tunneling and ballistic electron transport through ultrathin metal films to image inhomogeneities at buried metal–semiconductor interfaces (Bell and Kaiser, 1988; Kaiser and Bell, 1988). As a spectroscopic technique, BEEM provides the capability to measure local Schottky-barrier heights (SBH) and the electronic structure of such interfaces. BEEM operates in a three-terminal configuration, as depicted in Fig. 24. Electrons tunneling from the tip are injected with a current I_t into a thin (~ 100 Å) metal layer (base) that is deposited on a semiconductor surface (collector). We consider a case in which the base is grounded and the tip is biased at V_t. As the bias is swept, a threshold bias is reached when the collector current I_c becomes non-zero, and this represents the local SBH. These electrons experience ballis-

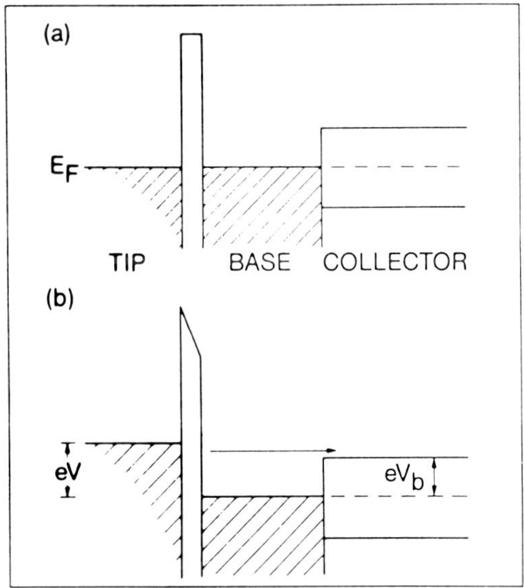

Figure 24. Three-terminal configuration of the BEEM experiment: (a) zero tunnel bias condition; (b) with a tunnel bias voltage above the barrier height (Kaiser and Bell, 1988).

tic transport, having traversed the thin metal layer without energy loss. By the simple theory, above the barrier threshold, the collector current varies linearly with V_t when I_t is the held fixed by feedback control of the tip–sample distance. In this mode, electronic states near the threshold and far above the threshold can also be characterized. It must be realized that, depending on the quality and thickness of the overlayer and the temperature of the system, electron–electron and electron–phonon scattering effects must be taken into account.

Since momentum and energy must be conserved at the ordered interface, the ballistic electrons are focused forwards providing a high degree of spatial resolution. Therefore regions of different barrier heights can be distinguished by scanning the tip at a bias above some nominal V_{SBH} and measuring I_c. Not only does such an image contain features due to SBH differences, but it may also contain geometric variations at the surface and at the interface. The surface topography can be handled by acquiring an STM image quasi-simultaneously, but the interface structure is more difficult to characterize.

The basic concept of BEEM can be described starting with free electron densities of states for the tip and the base, and considering a parabolic conduction band minimum (CBM) for the semiconductor. Electrons traversing the base with negligible scattering must conserve energy (E) and transverse momentum (k_t) at the interface where structural order exists. Satisfying both conditions requires that electrons that enter the semiconductor traverse the metal base with a small angular deviation from the normal (with respect to the interface plane) since the range of transverse momenta is limited. A critical angle, θ_c, with respect to the normal direction can be defined as

$$\sin^2\theta_c = \frac{m_t}{m}\,\frac{(V_t - V_{SBH})}{(E_F + V_t)} \tag{14}$$

where m is the free electron mass and m_t is the electron effective mass parallel to the interface.

The range of transverse momenta can be written in terms of a "transverse energy" E_t such that

$$E_t < \frac{m_t}{(m - m_t)}\,[E_x - E_F + e(V - V_b)] \tag{15}$$

where E_x is the energy associated with the component of the electron wavevector normal to the interface.

The small critical angle provides high spatial resolution. For example, a 100 Å Au layer on Si results in a critical angle of $\sim 4^\circ$ so that scanning the tip (the injector) can provide a mapping of inhomogeneities in the barrier height at the interface with $\sim$10 Å lateral resolution. The collector current can be written as

$$I_c(V) = R\,I_t\,C\,D(E_x)\,F(E)\,\mathrm{d}E_F\,\mathrm{d}E_x \tag{16}$$

with $E_{max} = E_t$ and $E_{min} = E_F - e(V - V_b)$, the minimum energy to overcome the Schottky-barrier (V_b is the base voltage). This yields a relation which is linear in injected current, I_t, and scattering resistance, R, which is taken as an energy-independent scattering parameter. Just above the threshold, $I_c \propto (V - V_b)^2$, and $I_c \propto (V - V_b)$ for larger V. The barrier height can be obtained by fitting the I_c versus V data.

Just as STS provides a means of measuring the surface electronic states, BEEM can be used to measure electronic states at the interface. Taking $\mathrm{d}I_c/\mathrm{d}V$ versus V_t, deviations from the apparent linear dependence of I_c versus V_t become clear and can be interpreted in terms of the interfacial band structure. The existence of multiple differential conductance thresholds implies that several conduction channels may exist at the metal–semiconductor interface. The first threshold is related to the barrier height and subsequent threshold values distinguish other aspects of the interface.

More complete theoretical treatments of BEEM take into account several factors that affect the collector current (Ludeke et al., 1991; Ludeke, 1993). In particular, the electron mean free path in the base results in an attenuation factor that depends exponentially on the electron energy; the transmission probability across the interface depends on both the degree of momentum conservation across the interface, due to the interface roughness or lack of epitaxy, and a quantum mechanical transmission factor across an abrupt potential step. Quasi-elastic scattering in the base, primarily from electron–electron scattering, results in a broadening of the momentum distribution incident upon the interface. In addition, temperature-dependent broadening must be included when fitting to data. Modeling all these effects has been shown to lead to an $I_c \propto (V - V_0)^{5/2}$ power dependence [rather than $I_c \propto (V - V_0)^2$) yielding a superior fit to BEEM spectroscopic data on interfaces where such considerations are important.

2.3 Instrumentation

The STM is a near-field imaging instrument operating on the principle of electron tunneling between an atomically-sharp probe tip and a sample surface. Images are obtained in the constant current or constant height modes by scanning the tip laterally over the surface while tunneling. The tunneling current, I_t, exhibits a nearly exponential dependence on the tip–sample separation, s; the tunneling probability is proportional to $e^{-2\alpha s}$, where $\alpha \sim 1\,\text{Å}^{-1}$ for a barrier height of ~ 4 eV. I_t is of the order of 1 nA for $s \approx 10$ Å, and is reduced by a factor of ~ 10 for every 1 Å change in separation. Maintaining a stable tunnel junction with an accuracy of ~ 0.01 Å, as the tip is scanned over the surface, is the most crucial aspect of STM operation and therefore demands that the apparatus exhibit *high mechanical stability*. In addition, since the tip and sample are typically mounted at points much further apart than the tip–sample separation, *thermal drift* must be compensated for and a high degree of *vibration/shock isolation* must be ensured. The STM head is typically designed to meet the criteria of the measurements which are to be performed, e.g. constant temperature versus variable temperature, atomic resolution versus wide scan range. The first step to meet these goals is to design a rigid structure in the STM head keeping uniformity of thermal expansion coefficients in the materials used. This provides the apparatus with a high resonance frequency, aiding in isolating vibrations from the surroundings; uniform materials provide good thermal drift characteristics. Obviously, tip structure and preparation is important to obtain atomic resolution images, and several methods have been developed and tested. One of the key technical problems in operating an STM is the *coarse approach* allowing the tip–sample separation to proceed from macroscopic distances to the tunneling regime. This aspect largely dictates the microscope design, being handled successfully in a number of ways – mechanically or electro-mechanically. *Microactuation* of the probe tip is usually accomplished using piezoelectric transducers for motion in each of the three orthogonal directions. With low-noise drive circuitry, electrical control can approach ≤ 0.01 Å. The electronics incorporate low-noise current preamplification, and, in the constant current mode, tunnel current feedback controls the height of the tip by adjusting the z-piezoelectric element as the tip is rastered in the (x, y)-plane. Scanning the tip, setting or scanning the tip–sample bias, and other data acquisition tasks are performed in real-time by a microcomputer, while image analysis and processing can be performed by the same microcomputer or off-loaded to workstations or mainframes. In this section, these basic elements of STM system design, as illustrated schematically in Fig. 25, are discussed.

2.3.1 Microscope Design: STM Heads

Positioning the sample in the tunneling regime typically begins several centimeters away using some *coarse approach* method. Once in the tunneling range, *fine motion* is typically accomplished by piezoelectric actuators. For various microscope designs, *coarse approach* has been performed by mechanical or non-mechanical methods, typically moving the sample towards the tip fixed on the piezoelectric actuator. If a purely mechanical coarse approach system is employed, the mechanical connection to the outside world must be decoupled from the microscope while scanning; the coarse advance steps must be small enough to accommodate the maximum extension of the z-piezo. Coarse approach can be accomplished electromechanically with a louse, as discussed previously, with piezoelectric-based slip-stick inertial mechanisms, or using piezoelectric inchworms.

The *fine motion* is used for final approach and scanning the tip during measurements. Several methods are illustrated in Fig. 26. Tip scanning was initially done using piezoelectric tripods, where elongation of each element provided x, y, z motion (Binnig et al., 1982a; Demuth et al., 1986; Sakurai et al., 1990). Soon thereafter, segmented tubular piezoelectric elements utilizing elongation for z motion and bending for x, y motion (Binnig and Smith, 1986; Lyding et al., 1988; Snyder and de Lozanne, 1988), bimorph-driven tripods (Blackford et al., 1987), or microfabricated bimorphs (Akamine et al., 1989) were introduced. Smaller and lighter elements increase the resonance frequency of the elements while reducing the effects of thermal gradients. Cross-talk between orthogonal motions are minimal for scanning ranges that extend up to the micron scale. Sensi-

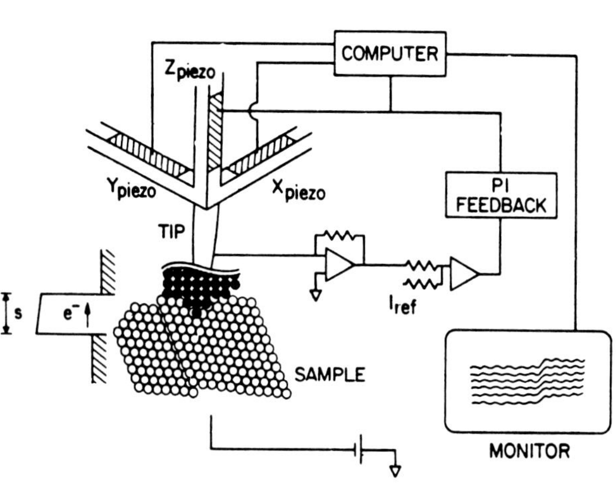

Figure 25. Schematic diagram indicating the major components of an STM system. An atomically-sharp tip is biased with respect to the sample. In the constant current mode, the three-dimensional piezoelectric scanner assembly is scanned over the surface by the computer which controls the x-piezo and y-piezo biases. As the tip is scanned, the amplified tunneling current is compared to a reference signal, and the result is fed back as an error signal to control the bias on the z-piezo. The error signal also provides the height signal to the computer to form the image shown on the display monitor (Kuk and Silverman, 1989).

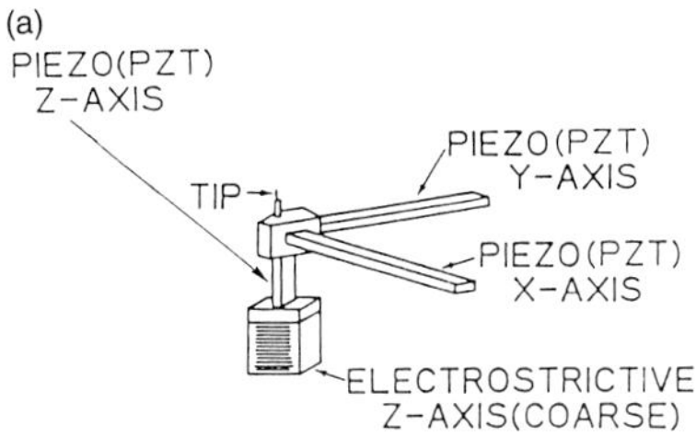

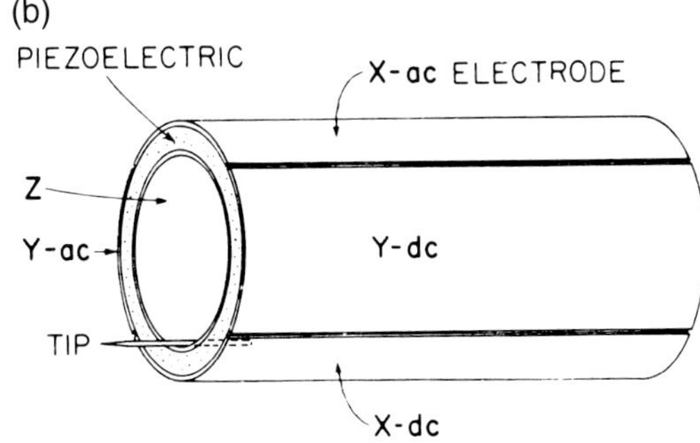

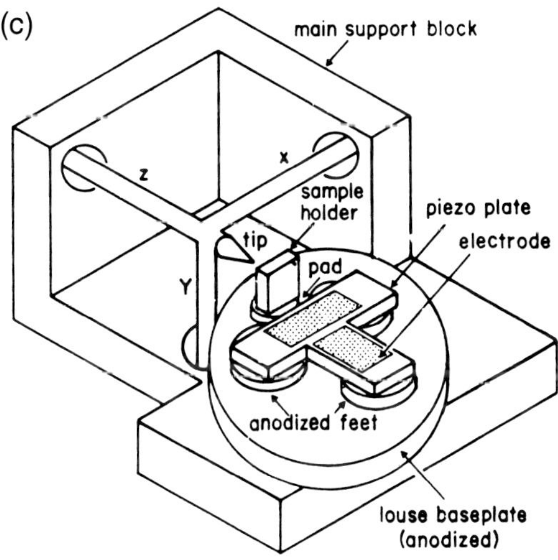

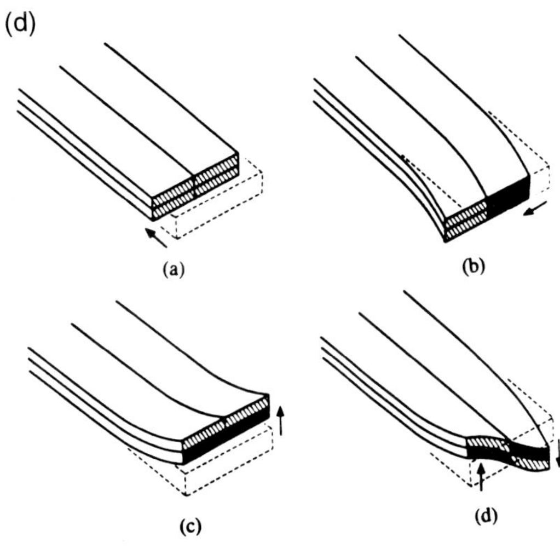

Figure 26. (a) A piezoelectric tripod provides three-dimensional fine positioning of the tip while an electrostrictive element allows coarse approach (Sakurai et al., 1990). (b) A quartered tubular piezoelectric element provides, *x*, *y*, and *z* motion; in this case the tip is mounted off-axis (Binnig and Smith, 1986). (c) Bimorph-elements (hidden in block) mounted on the main support block drive the tripod frame (Blackford et al., 1987). (d) A microfabricated piezoelectric actuator permits lengthwise, lateral bending, vertical bending, and twist motions (Akamine et al., 1989).

tivities of the order of tens of angstroms per volt make motion in steps of 0.1 Å in the *x*- or *y*-direction easily controlled by independent operational amplifiers each connected to a digital-to-analog converter for raster scanning by the data acquisition microcomputer.

We noted in the introduction the elaborate set-up for the first STM from Binnig and Rohrer using magnetic levitation, a louse, and a tripod assembly (Binnig et al., 1982b, 1983b). In this section, several other designs are discussed to illustrate the conceptual variations in design and the evolution of the instrumentation.

The UHV-STM design of Sakurai et al. (Sakurai et al., 1990), shown in Fig. 27, was modeled on the Demuth design (Demuth et al., 1986). The sample is mounted on a double-hinged flip mechanism so that the sample approaches the tip from a vertical position rapidly until (right) the sample arm swings down (on pivot A) and rests on the stopper (foot) located near the tip and slightly protruding. The swing arm allows the sample to be accessed by other surface probes such as low-energy electron diffraction (LEED) and Auger electron spectros-

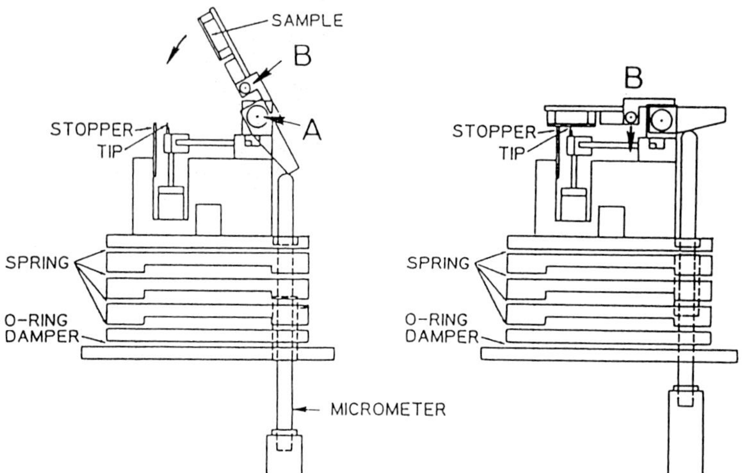

Figure 27. The UHV-STM with a double-hinged sample approach allows sample approach with hinge A and fine mechanical approach with hinge B (Sakurai et al., 1990).

copy (AES), and the tip can be viewed by a moveable field ion microscope. Once the sample rests on the stopper, two coarse-approach philosophies are employed. First, exploiting the mechanical advantage between the second pivot (B) and the stopper ($\sim 40{:}1$), a fine-thread screw tilts the sample arm to bring the sample into the tunneling range. Secondly, with the feedback activated, the tip can be extended by the electrostriction actuator (motion $\sim 12\ \mu m$) as well as by the z-piezoactuator until a tunnel current is sensed. Once tunneling is achieved, the mechanical drive is decoupled for vibration isolation, the electrostriction actuator is held fixed, and the fine control takes over operation of the tip.

One of the first designs for imaging and spectroscopy at low temperatures is shown in Fig. 28 (Fein et al., 1987). Used for applications in superconducting materials, low temperatures are obtained by insertion into a research liquid helium dewar and the unit is surrounded by a superconducting magnet, so the design must be compact. In this design, materials were chosen to minimize differential expansion or contraction to obtain a satisfactory coarse approach without tip crashes as the unit is cycled between room temperature and liquid helium temperature; molybdenum was used for the majority of parts for its low coefficient of thermal expansion. The coarse approach is performed before the

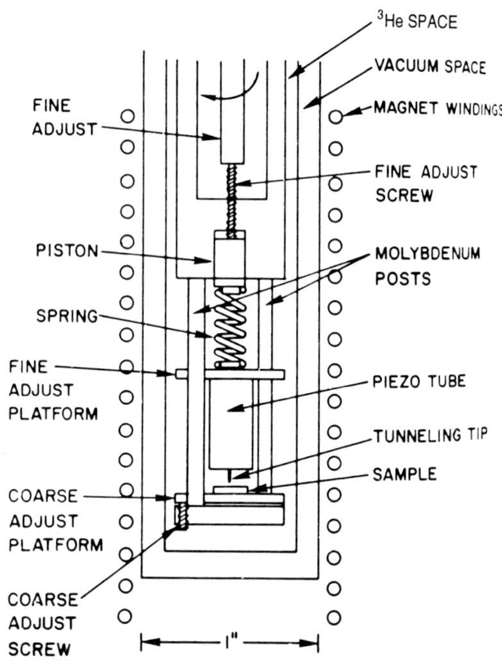

Figure 28. STM designed for use in a liquid helium dewar in high magnetic fields incorporating mechanical coarse approach and a tube scanner (Fein et al., 1987).

sample is introduced into the dewar by simply deflecting the cantilever sample mount. The fine approach is done mechanically by distorting the thin plate on which the tubular piezo is mounted using a fine-pitch screw which pushes on a soft spring in contact with the plate. This "differential spring" produces a mechanical advantage of ~100:1 giving sufficient sensitivity to enter into the tunneling regime.

This microscope uses a segmented (quartered) tubular piezoelectric element introduced by Binnig to replace the tripod as a compact element with a significantly higher resonance frequency (as high as ~100 kHz) (Binnig and Smith, 1986). The inner surface of the tube is usually kept at a uniform potential and independent potentials are applied to each of the four outer quadrants. Operating in a bipolar configuration, opposite polarities are applied to opposite quarters to provide parallel motion, as illustrated in Figure 26b. The electromechanical properties for bipolar operation of the tubular piezo are based on the piezoelectric coefficient, d_{31}, and geometric analysis (Chen, 1992). The z-motion is obtained by establishing a uniform potential difference, ΔV, between the inner and outer electrodes, i.e.,

$$\Delta z = d_{31} \frac{\Delta V L}{h} = K_z \Delta V \qquad (17)$$

where L is the length of the element, h is the wall thickness of the tube, and K is the piezo constant. Lateral motion of the tip is accomplished by applying opposite voltages on opposed quarters, so that one side of the element contracts and the other expands creating positive and negative stresses parallel to the tube which translate into strain perpendicular to the tube axis. Analysis of this situation leads to

$$\Delta x = 2\sqrt{2}\, d_{31} \frac{\Delta V L^2}{\pi D h} = K_{\parallel} \Delta V \qquad (18)$$

where D is the diameter of the element. The expression for Δy is identical. In practice, three-dimensional motion of the tip is obtained by modulating the center contact with respect to the potentials on the quadrant to obtain z-motion of the tip while the quadrants are modulated with bipolar x and y potentials, for raster scanning. For a PZT-5H element, where $d_{31} = 2.74$ Å/V, and with $D = 1.27$ cm, $h = 0.102$ cm, and $L = 1.27$ cm, $K_z \approx 34.1$ Å/V and $K_{\parallel} \approx 30.7$ Å/V. In order to maintain control to < 0.1 Å, noise in the driving voltages must be kept below 3 mV.

With the proper piezoelectric materials and dimensions and increased voltages, scanning ranges up to tens of microns can be achieved. Intermediate translators may be used to scan widely-spaced regions on large-area samples (Hipps et al., 1990).

The microscope design shown in Fig. 29 uses coaxial tubular piezoelectric elements, employs a coarse slip-stick inertial approach, and exhibits excellent temperature compensation resulting from a single mounting point for the coaxial tubes (Lyding et al., 1988). The outer tube, on which the sample support is mounted, is used as a sample approach by providing a series of pulse sequences involving pulsed outward motions, where the sample remains stationary as the force of static friction is exceeded, and slow inward motions pull the sample nearer the tip. The sample is removed using the opposite sequence of pulses. Adjusting the magnitude and timing of the pulses can vary each step length from ~5 Å to ~1 μm, and the speed of approach up to ~1 mm/s. The inner quartered tube operates in the same fashion as the instrument described previously. In the original design the microscope was housed in a temperature control shroud, which provided an ability to heat or cool with the microscope maintained at a fairly uniform

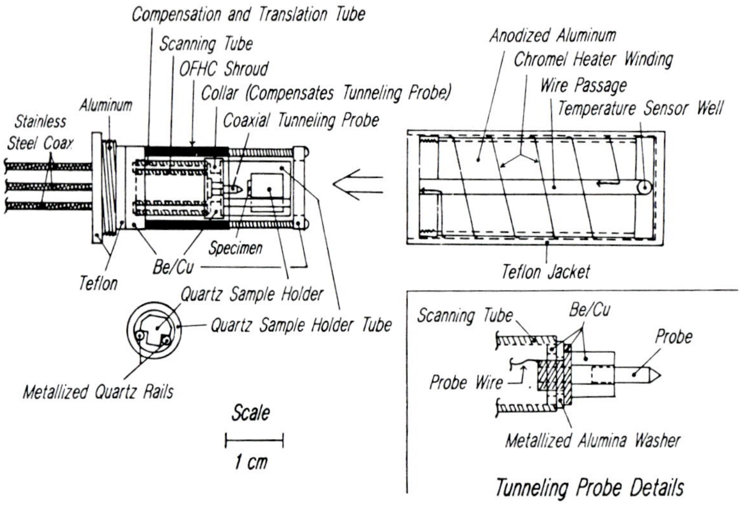

Figure 29. STM designed with inertial coarse approach and coaxial piezoelectric tubes for temperature-compensation. The STM is operated in a thermal shroud (Lyding et al., 1988).

temperature. As previously noted, the variable temperature capability of this microscope is enhanced by the fact that both tubes, being constructed of similar equal materials, expand and contract at the same rate with respect to their common mounting point.

The STM illustrated in Fig. 30 (Frohn et al., 1989) has been operated in ultrahigh vacuum (UHV) at variable temperatures down to ~ 20 K. The high thermal conduction to the sample is made possible by

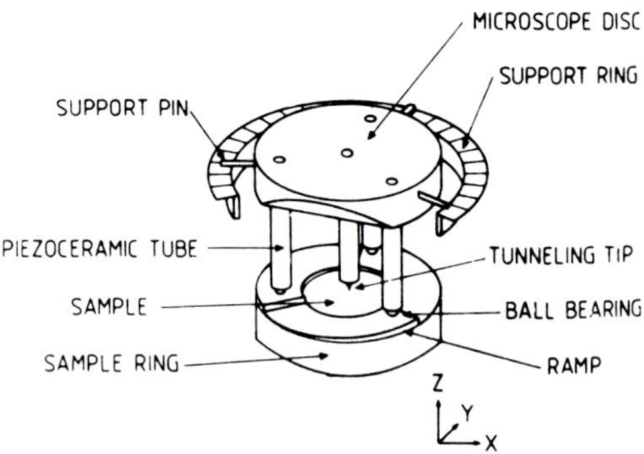

Figure 30. STM using a novel coarse approach philosophy in which three piezoelectric support legs are inertially-driven to rotate the microscope disc down three ramps on the sample ring (Frohn et al., 1989).

its direct attachment to a liquid He dewa This microscope design concept is ba on a compact design using a novel proach mechanism (Besocke, 1987) tip holder and three piezoelectric mounted as a unit which rests on a graded ring that surrounded the sample. The tip unit is set on the ring from above and is decoupled from the manipulator. Applying pulses on the three legs in a tangential direction allows the tip unit to rotate and descend down the grades until the tunneling regime is reached. At that point lateral movement of the legs can be used for x-y scanning. The sample can be heated or cooled by attachment to a heat source or sink, while the small size and low thermal conductivity of the tip assembly keeps thermal gradients within the microscope to a minimum (Wolf et al., 1991).

High-temperature STM operation is especially problematic due to thermal drift, as well as due to the relatively low Curie temperature of the piezoelectric elements. A recently introduced high-temperature UHV STM (Watanabe et al., 1990) utilizes direct sample heating and exhibits excellent thermal isolation of the piezoelectric elements from the heater. This is accomplished by mounting a small sample on a pyrolytic BN insulating film on top of a graphite heater. The anisotropic thermal conductivity of BN ensures a uniform temperature distribution, and the direct resistance heating requires only small wires to attain temperatures up to $> 400\,^{\circ}\mathrm{C}$. The sample is mounted on an inchworm for coarse positioning. The tip is mounted on a block which is driven by four Invar arms each attached to a bimorph piezoelectric disc. The piezoelectric elements remain near room temperature for the entire range of sample temperatures so that they maintain a constant sensitivity. Furthermore, the symmetric arrangement of drive arms compensates for thermal drift effects.

Construction of a microfabricated STM of sub-millimeter dimensions has been demonstrated using planar processing techniques (Akamine et al., 1989), as illustrated in Fig. 31. Besides opening up new applications in lithography, data storage, or sensor technology, the small size reduces the magnitude of thermal drifts, increases mechanical resonance frequencies, and reduces sensitivity to external vibrations. In this design, the cantilever of dimensions $1000 \times 200 \times 8$ μm was constructed from alternating layers of metal electrodes, dielectric films, and well-oriented piezoelectric ZnO films on an Si substrate to provide full three-dimensional motion. Tips were formed at the end by metal evaporation or attachment of metal wires or particles.

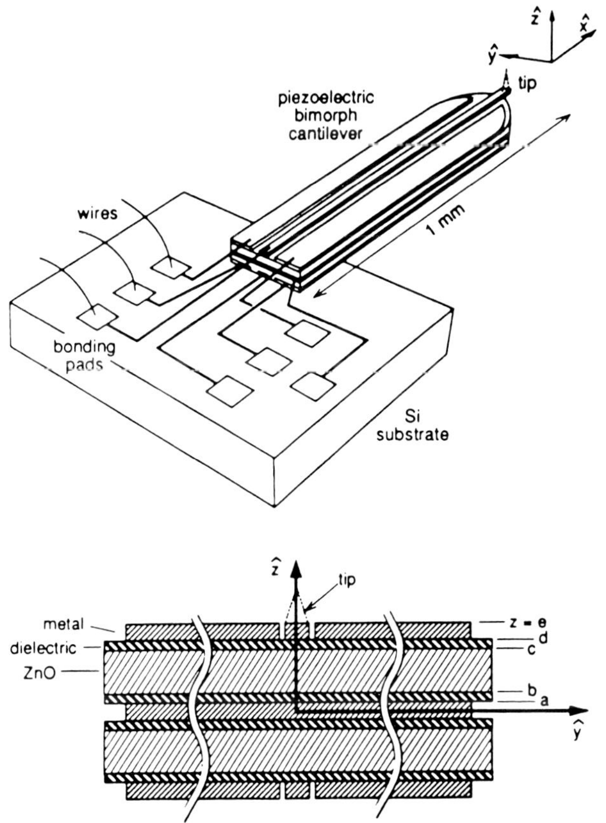

Figure 31. Design of a microfabricated STM using a piezoelectric bimorph cantilever (section) for three-dimensional motion (Akamine et al., 1989).

2.3.2 Tips

Theory suggests that the lateral resolution that can be obtained by an STM is directly related to the atomic sharpness of the tip. The corrugation amplitude, Δ, is given by the relation

$$\Delta \propto e^{-\beta(R+d)} \tag{19}$$

where $\beta \sim \varkappa^{-1} G^2/4$, with $\varkappa$ the inverse decay length, G the smallest surface reciprocal lattice wavevector (directly related to the corrugation width), d the gap distance, and R the radius of curvature of the tip (Kuk and Silverman, 1986). That tips can be prepared with one or a few atoms protruding and the fact that the tunneling probability decays rapidly with separation permits atomic resolution to be achieved more or less routinely, although multiple protrusions in approximately the same plane can produce anomalous images. Several materials have been used successfully, with specific methods of tip preparation having been developed for each material, ranging from mechanical extrusion and/or grinding (e.g., PtIr tips) to electrochemical etching (e.g., W tips). In addition, microtips of sub-micron dimensions have been fabricated by e-beam deposition to obtain a high aspect ratio for the study of deep structures (Akama et al., 1989). It must be noted that the material at the surface of the tip is important since spectroscopy depends on the joint density of states between the tip and the sample.

Examination of the tip by electron microscopy gives a general measure of shape and sharpness, but not an atomic scale view. Atomically-resolved STM images are generally obtained after in situ preparation using high-biases (field-evaporation or atom transfer/rearrangement) or high currents (tip annealing), but care must be taken not to blunt the tip by such processing.

While typically routine methods are followed to fabricate tips that operate satisfactorily, one pioneering study established a connection between the tip atomic structure of W, as obtained by field ion microscopy (FIM), and STM imaging resolution (Kuk and Silverman, 1986). Tungsten is a tip material which has received widespread use, partly because the surface energy of the (110) facets of b.c.c. metals can be formed by annealing in a high electric field. Using electrochemically-etched single-crystal W(100) wire and in-situ field-evaporation, a tip with six atoms protruding as a distorted $p(2\times2)$ reconstructed W(100) surface was characterized prior to and after imaging the Au(001)-(5 × 1) reconstruction, as shown in Fig. 32. The measured STM imaging resolution of ~4 Å is in agreement with theory. Smaller tip clusters were found to be unstable, although they are absolutely essential for imaging close-packed metal surfaces. The sharpness of the W tip was severely reduced, as was the measured corrugation of the Au surface using such a tip after cleaning using field evaporation.

It should also be noted that spin-polarized imaging on the nanometer scale has been demonstrated using ferromagnetic CrO_2 tips is study Cr(001) (Wiesendanger et al., 1990) and ferromagnetic Fe tips to study magnetite (Fe_3O_4) (Wiesendanger et al., 1992).

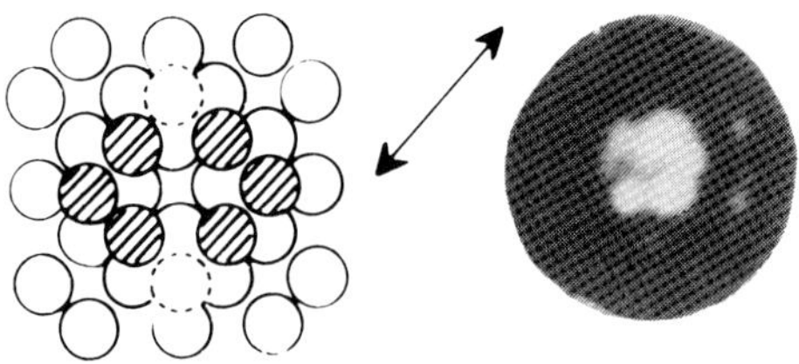

Figure 32. The FIM image of a tip grown on a single-crystal tungsten wire on the right is interpreted as originating from the arrangement of six tungsten atoms (left) in a b.c.c. (110)-like structure (Kuk and Silverman, 1986).

2.3.3 Vibration and Shock Isolation

The first STM was levitated on a superconducting magnet system (Binnig et al., 1982b). This extremely effective vibration isolation system demonstrated and developed the true potential of the technique (Binnig and Rohrer, 1987), permitting Binnig and Rohrer to concentrate on understanding the tunneling mechanisms and spatial resolution of the instrument. All of these unknown aspects combined to prevent previous researchers from achieving vacuum tunneling, but vibrational isolation was the most serious. With better understanding of the tunneling mechanism and the experimental constraints, relaxing the sophistication of the vibration isolation system and refining simpler designs has led to the proliferation of STMs of both custom and commercial types.

Isolation of the STM from vibration and shock from the external environment is critical for its operation. Once the STM head design has been selected, the vibration isolation system to operate in the ambient environment can take several forms. The elements of a vibration isolation system are depicted in Fig. 33. The principal consideration in planning the vibration isolation system is the high sensitivity of the current to tunneling distance ($\Delta I/I$ per angstrom is ~ 10) and the magnitude of topographic variations in the z-direction (~ 0.1 Å or less). The goal is to limit the vibration amplitude between the tip and the sample to ≤ 0.01 Å. Typical sources of vibration are from the building floor (< 5 Hz); from electromechanical devices, ventilation systems, and transformers (typically < 120 Hz, with particular activity in multiples/sub-multiples of the line frequency); and from acoustic noise. The STM can be isolated from high frequency vibration (> 1 kHz) by a factor of $\sim 10^6$

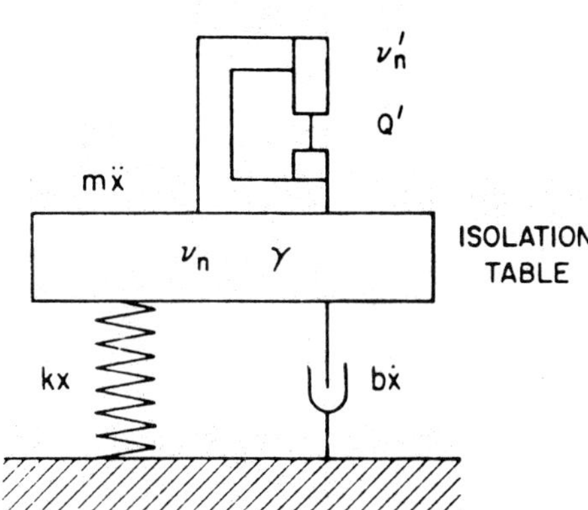

Figure 33. Block diagram showing parameters for the modeling of vibrational isolation for a typical STM (Kuk and Silverman, 1989).

using spring suspensions or elastomer supports. Fabricating the STM as a rigid structure with a resonance frequency in the high kHz range is the first step in decoupling the tunneling gap from low frequency vibrations. However, the suspension systems themselves exhibit low frequency resonances; their amplitudes can be reduced (while also increasing the slope of the high-frequency roll-off) by the use of multiple suspension stages and by mounting the entire apparatus on pneumatic supports. Disturbances due to shock can be removed by incorporating damping in the vibration isolation system.

A practical anti-vibration system consists of an STM chamber suspended on pneumatic legs or hanging from a spring suspension. The STM is itself structurally rigid, and it rests in the vacuum system on a series of stacked plates separated by elastomer spacers (Gerber et al., 1986). Recent calculations have found that optimum performance can be obtained by using a few stacked plates for STM mounting with soft suspension of the entire chamber (Schmid

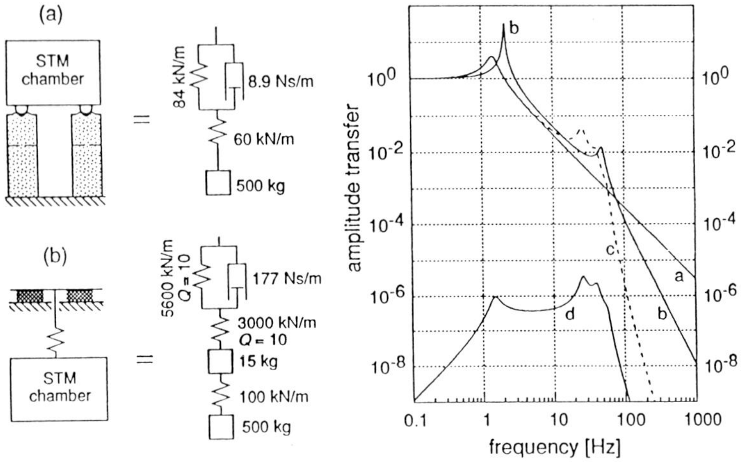

Figure 34. Amplitude transfer functions for: (a) chamber supported by a pneumatic suspension system; (b) chamber supported by a spring (with damping); (c) STM supported on a three-stage stack in a pneumatically supported chamber (not illustrated); (d) tip–sample distance of STM with resonance frequency of ~3 kHz (Schmid and Varga, 1992).

and Varga, 1992). Figure 34 depicts amplitude transfer functions for a chamber supported by (a) a pneumatic suspension system or (b) a spring suspension system with damping and incorporating realistic values of spring constants and damping coefficients. It is seen that the spring suspension has a higher frequency but higher amplitude resonance with a steeper high frequency roll-off than the pneumatic system. However, the pneumatic system is more practical. When the STM is supported on a three-stage stack (c) in a chamber suspended by pneumatic legs, the high-frequency roll-off is greatly enhanced. Furthermore, if the STM with resonance is at ~3 kHz, the reduction in coupling to the tunneling gap over the entire frequency range is approximately six order of magnitude or better, as shown in curve (d).

An alternative method of damping involves suspending the STM from coil springs and using an eddy-current damping system, as depicted in Fig. 35 (Okano et al., 1987). This two-stage caged system involves hanging a middle frame (II) from the base frame (I), which may be the chamber possibly sitting on a pneumatic suspension, by three springs. The exterior frame (III) is likewise supported by three springs attached to II. Each coil spring end is terminated by an elastomer ring. One eddy current damper is shown; it consists of copper blocks (IV) and (VI) surrounding a permanent magnet. These dampers attenuate vertical and rotational motion of the suspension. The transfer function for this suspension system, calculated and then compared experimentally using a vibration transducer and an acceleration sensor, is shown in Fig. 35 B. Without damping, there are two main resonances at ~2 Hz and ~5 Hz and an acoustic noise peak at ~50 Hz. The inset shows the vibration of the set-up as a result of mechanical shock. Introducing damping suppresses all the vibrational features, and shocks are rapidly damped.

2.3.4 Electronics

Figure 36 depicts a typical schematic diagram of analog control electronics for STM operation. Note that in this diagram (Kuk and Silverman, 1989) there are two *z*-piezoactuators – one microcomputer-controlled and one feedback-controlled. Scanning is accomplished by microcom-

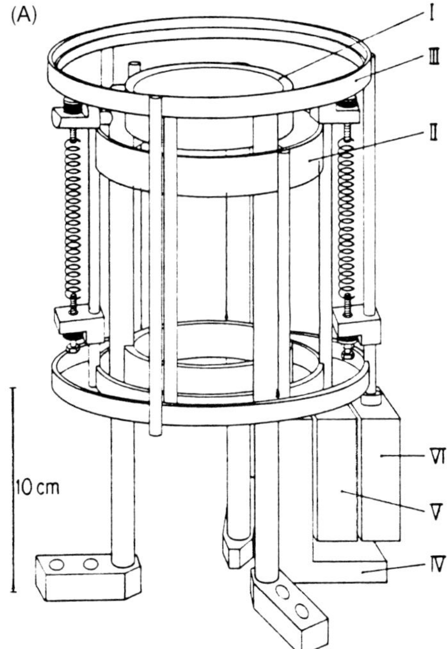

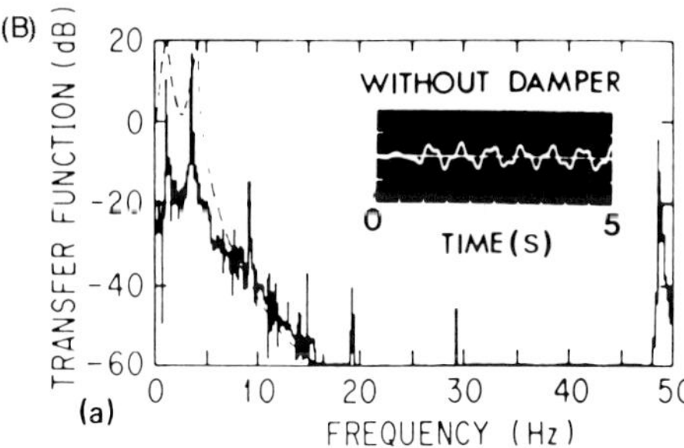

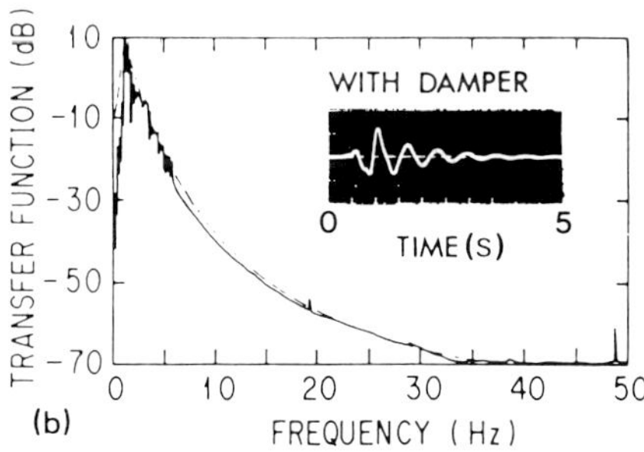

Figure 35. (A) STM suspended by coil springs with eddy-current damping in a two-stage caged system. The calculated transfer function is compared with experiment (B): (a) without damping; (b) with damping. The oscilloscope traces show the response of the suspension system to externally-applied shock (Okano et al., 1987).

puter-controlled digital-to-analog converters (DACs) operating high voltage preamps connected to the x (DAC2) and y (DAC3) piezoactuators while the z piezoactuator (DAC4) sets a coarse z-position. For single-tube piezo scanners operated in the bipolar mode, the z-bias is connected to the inner cylinder while the $\pm V_x$ and $\pm V_y$ are applied to opposite quarters of the outer type. The tip is biased by a variable voltage which is also microcomputer-controlled (DAC1); it is important that the tip be shielded from both leakage currents and capacitive coupling to the modulated piezo biases. The tunneling current is monitored through the sample, which is connected to a current preamplifier. Information passes to the microcomputer through analog-to-digital converters (ADCs).

Three modes of operation can be distinguished – constant height, constant current and spectroscopic. In the *constant height mode*, the tip–sample distance is set and the tip is scanned while the microcomputer acquires tunneling current as a function of position through the *current* ADC. In the *constant current mode*, the preamplifier signal is passed through a logarithmic amplifier to linearize tunnel current and tip–sample separation (i.e., $\log I \propto s$) and this signal is compared with a voltage related to the desired tunneling current (*demanding current* or reference current) using a comparator. The error signal generated by the feedback circuit is processed through both a proportional amplifier and an integrating amplifier and then passed through a high-voltage amplifier to correct the tip position to obtain the desired tunneling current. This error signal is likewise read by the microcomputer (*tip position*, ADC) and is directly translated to the height.

Analog feedback has often been used in STM control systems, but digital feedback has recently been introduced in several in-

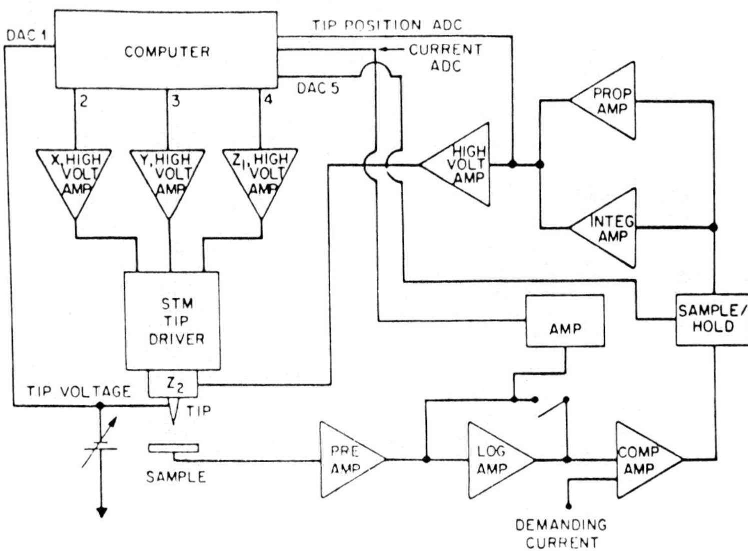

Figure 36. Schematic diagram of STM control electronics illustrating circuits for constant-current, constant-height, or spectroscopic modes (Kuk and Silverman, 1989).

struments. In studying the details of the feedback system, including the electronics and piezoactuators, the response of the scanners introduces an inherent time delay in the circuit that must be addressed. In practice, the feedback loop is operated with the maximum response which does not cause the system to oscillate; variable *RC* filtering, which introduces an additional phase shift, is used to prevent instabilities due to sudden changes in tip current due to irregularities in the surface or externally-introduced shock. Operating with too high a time constant limits the feedback response, resulting in unacceptable smoothing of the surface image.

Several *spectroscopic modes* exist that utilize a sample–hold to open the feedback loop during $I-V$ measurements. While sampling, the height is set at a desired tip–sample bias and tunneling current; then, while the height is held fixed, the bias is scanned and the current versus bias is recorded through the *current* ADC. Thus an image and a spectrum are obtained for various positions.

Combining lock-in techniques with the STM electronics can significantly reduce noise (Abraham et al., 1988), and such techniques are commonly used in non-contact AFM systems, as will be discussed in the next section. Briefly, a dither voltage is placed on the z-piezo to oscillate the tip in and out at a specific frequency. The lock-in synchronously detects the signal, reducing contributions from unwanted noise with frequency components spanning the rest of the spectrum.

Progressively increasing speeds in communication between microcomputers and instrumentation have allowed the introduction of real-time digital feedback and digital signal processing for STM applications. Digital feedback control is advantageous since it permits a complex signal processing algorithm to be implemented in the software. For example, one such system contains two software control modules (Morgan and Stupian, 1991). One module emulates the logarithmic amplifier using a look-up table to linearize the tip–sample distance versus the current relation. The second module is a proportional-integral-differential controller, which allows the gain and the feedback loop parameters to be adjusted during a scan to achieve optimum operation, depending on tip and surface conditions.

2.3.5 Microcomputer Control

Figure 37 shows a microcomputer-based STM workstation for data acquisition and image processing. Control is implemented by a microcomputer and software operating DACs, ADCs, and video peripherals and driving various image analysis and spectral analysis routines. Typically, all control functions are controlled from the keyboard, and images are displayed during acquisition. After acquisition, several functions are generally available to noise-filter, analyze, and portray data in the form of topographic and spectroscopic images and atomically resolved spectra.

2.4 Semiconductor Surfaces

2.4.1 Si(111)

Si(111)-7 × 7 – Atom-Scale Topography and Spectroscopy

The use of STM to solve the geometric structure (Binnig et al., 1983) and the electronic structure (Hamers et al., 1986) of the Si(111)-(7 × 7) reconstruction has briefly been discussed in Sect. 2.2. Such work added crucial detail to support the dimer-adatom-stacking-fault (DAS) model proposed independently, based on an electron microscopy analysis (Takayanagi et al., 1985). Total energy calculations further supporting the DAS model have recently been reported (Brommer et al., 1992; Stich et al., 1992). Other Si(111) structures exist, most notably the (2 × 1) which represents a doubly periodic superstructure in one direction, while the (7 × 7), the stable reconstructed phase, is sevenfold-periodic along both principle axes spanning 49 surface unit cells. In addition, (5 × 5) and (9 × 9) reconstructions have been observed (Becker et al., 1986; Becker et al., 1989; Feenstra and Lutz, 1990; Feenstra and Lutz, 1991 a).

The first in vacuo experiment of Binnig and Rohrer unambiguously showed that the (7 × 7) structure comprises an array of 12 Si adatoms, which were found to be arranged locally in a 2 × 2 structure, nine dimers on the sides of triangular subunits of the 7 × 7 unit cell, and a stacking fault layer (Takayanagi et al., 1985). These STM topographic images, shown in Fig. 12 a, are line scans recorded on a chart recorder with $V_t = 2.9$ V. The measured rhombohedral unit cell diagonals of 46 × 29 Å bounded by deep corner holes

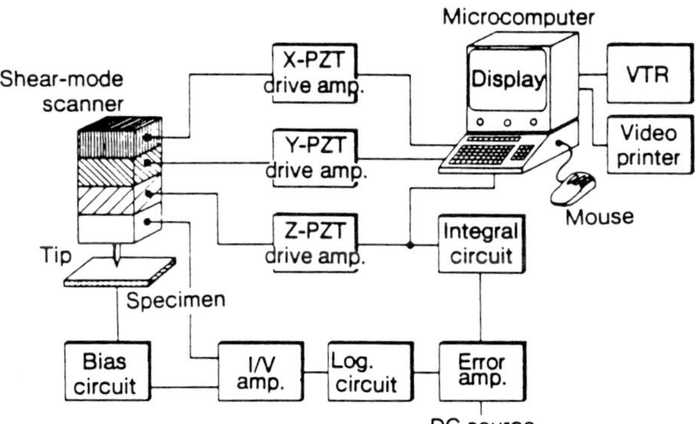

Figure 37. Microcomputer-based STM workstation with video accessories as peripherals (Iwatsuki and Kitamura, 1990).

encompass twelve maxima. Since tunneling mainly involves dangling bonds protruding into the vacuum, a model incorporating twelve adatoms was supported. That is, the twelve maxima are from threefold-coordinated Si adatoms with empty sites at the corners (corner holes). The "rest atoms", atoms in the next layer with unsatisfied dangling bonds, were not imaged at this bias. In addition to verifying an adatom model, a displacement of the adatom structure with respect to the underlying surface by ~2 Å for half of the unit cell was observed for positive sample bias (unoccupied states) due to the stacking fault.

Further details of the reconstruction were explained by subsequent studies using spectroscopic imaging (Becker et al., 1985b) and CITS (Hamers et al., 1986c). As noted in Sect. 2.2, such studies represented the first demonstration of the need for combined spectroscopic *and* topographic imaging as a means to explain STM images. Figure 21 presented CITS images at three bias voltages mapping occupied states. A more complete set of CITS images is displayed in Fig. 38, showing that the adatoms have both occupied and unoccupied character (−0.35 eV and 0.50 eV). The 0.35 eV state gives rise to the faulted/unfaulted asymmetry typically observed in unoccupied state images. Another unoccupied state appears between the adatoms with different onset energies for the faulted and unfaulted halves.

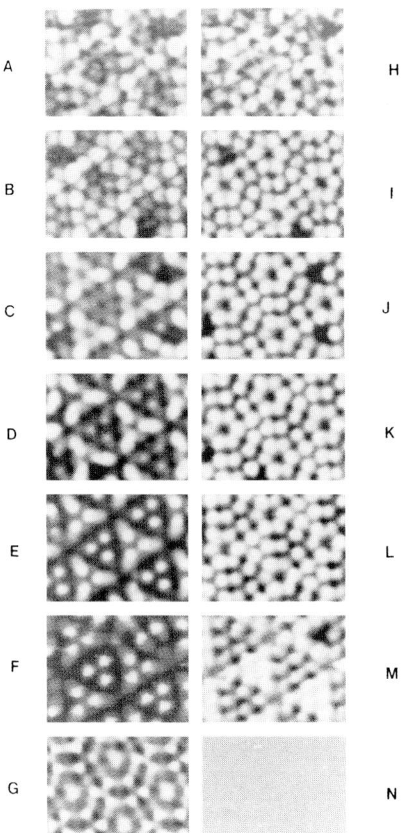

Figure 38. CITS images of the Si(111)-7 × 7 surface with the feedback loop at $V_t = -2$ V. Images are shown at the following biases: (A) +0.15 V, (B) +0.25 V, (C) +0.65 V, (D) +0.75 V, (E) +0.95 V, (F) +1.55 V, (G) +2.0 V, (H) −0.15 V, (I) −0.25 V, (J) −0.65 V, (K) −0.75 V, (L) −0.95 V, (M) −1.55 V, (N) −2.0 V (Hamers et al., 1986b).

Si(111)-2 × 1 – Structure, Spectroscopy, and Local Defect Structure

An analysis of the Si(111)-2 × 1 reconstruction on cleaved Si(111) samples (Feenstra et al., 1986) supported the proposed π-bonded chain structural model (Pandey, 1981; Pandey, 1982) of this reconstruction. In the 2 × 1 reconstruction, occupied and unoccupied surface states around the surface Brillouin zone (SBZ) boundary are separated by ~0.4 eV, while states at the SBZ center mix with the bulk bands. The existence of states within the gap provided the contrast for surface images of the π-bonded chains shown in Fig. 39. Long atomic rows along [01$\bar{1}$], separated by 6.9 Å, are observed to coexist with disordered regions where the rows merge and hillocks are observed. This spacing is consistent with delocalized π-derived bands

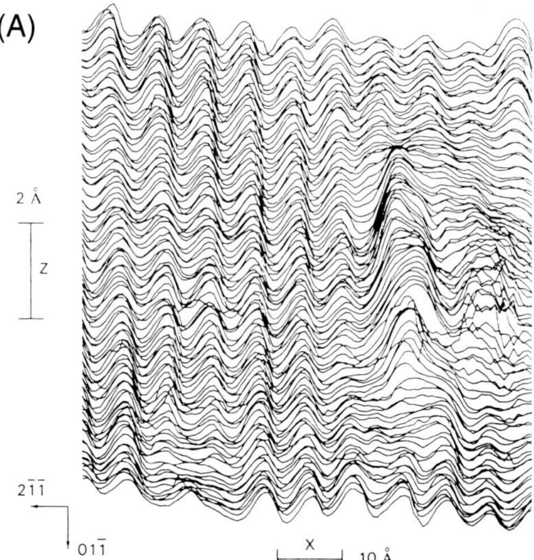

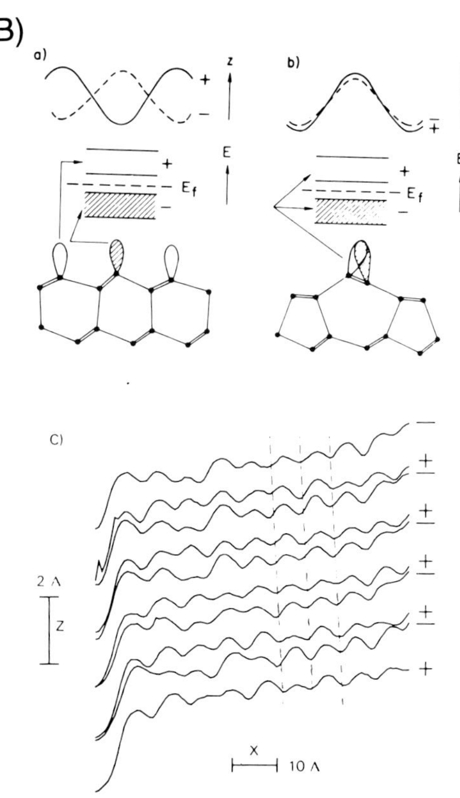

Figure 39. (A) STM image of cleaved Si(111) surface acquired with $V_t = -0.6$ V. 2×1 π-bonded chains are observed on the left side and a disordered region is seen on the right side. (B) Expected corrugation for tunneling into and out of surface states of the (a) buckled model and (b) π-bonded chain with experimental results (c) that indicate sample biases switched to +0.8 V (+) and to −0.8 V (−) on successive scans (Feenstra et al., 1986).

along the chains. Line scans across the chains produced a measured corrugation amplitude of 0.54 Å and the in-phase corrugation for both polarities is also consistent with the π-bonded chain model. Subsequent spectroscopic measurements of the 2×1 surface were performed (Stroscio et al., 1986). The surface states with maxima at the zone-center (±0.2 eV) and at the zone edge (±1.2 eV) were observed in $(\mathrm{d}I/\mathrm{d}V)/(I/V)$ spectra obtained at various heights above the surface chains. Differences in the electronic structure due to structural defects were also observed (Feenstra et al., 1986). For biases below those used to image the ordered 2×1 structure, the hillocks were clearly imaged, indicating disorder-induced modification of the electronic structure; these observations supported results of EELS measurements of the 2×1 surface band gap (DiNardo et al., 1985).

Phase Transitions Between Si(111) Structures

The kinetics of phase transitions between ordered reconstructions on Si(111) up to the high temperature (1×1) structure have been obtained by temperature-dependent STM (Feenstra and Lutz, 1991a; Kitamura et al., 1991; Miki et al., 1992; Hibino et al., 1993). Similar measurements on Ge(111) have also been reported (Feenstra et al., 1991).

In an analysis of the time- and temperature-dependence of the $(2 \times 1) \Rightarrow (5 \times 5)$, (7×7) reconstructions on an STM designed to operate at temperatures up to $\sim 400\,°C$ (Feenstra and Lutz, 1991 a), it was found that the process is temporally nonlinear for $T \leq 330\,°C$. This is due to the fact that the transition is dominated by the number of surface adatoms available; at $\sim 320\,°C$, only an initial conversion to an adatom-covered 1×1 surface is required to make the adatoms available (another source of adatoms at low temperatures is at 2×1 domain boundaries). At $\sim 330\,°C$, the surface orders into the (5×5) and (7×7) by creating DAS structures, thereby liberating more atoms for further growth, as shown in Fig. 40. In this image, monolayer-deep "holes" also appear; these are additional sources of adatoms. Above $\sim 330\,°C$, (5×5) structures predominate with some (7×7) structures, forming without the adatom-covered 1×1 intermediate. Ordered 5×5 "fingers" extending along the $[01\bar{1}]$ direction and nucleating around domain boundaries of the original (2×1) coexist with monolayer-deep "holes". Here

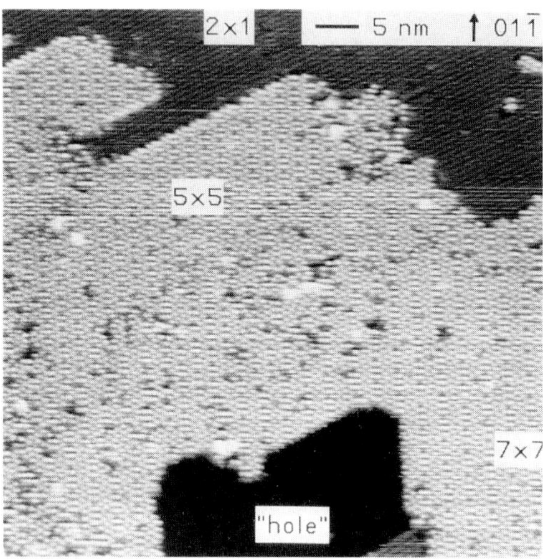

Figure 40. STM image of the Si(111) surface annealed at 330 °C for 900 s. (2×1), (5×5), and (7×7) domains are seen with monolayer-deep holes containing (5×5) and (7×7) structures (Feenstra and Lutz, 1991 a).

the kinetics make it possible to form the structures so long as the number of atoms is conserved. In particular, the (7×7) is not favored because of a deficiency of adatoms; the 2×1 and 5×5 contain the same number of atoms while the 7×7 contains 4% more atoms.

The formation of the (7×7) from the high-temperature (1×1) has been imaged with a new UHV microscope capable of stable operation above 900 °C (Iwatsuki and Kitamura, 1990). Figure 41 begins with a current image acquired at 880 °C. Due to the response of the feedback mechanism to avoid tip crashes at steps, steps bands appear dark if the left-hand terrace is higher than the right-hand terrace and appear light in the opposite circumstance. Lowering the temperature to 860 °C results in completion of the (7×7) on the upper side of the step. The atoms appear to be in motion near the lower side of the step because the step itself is still in motion. At 840 °C, the entire surface is covered with the (7×7) and the step is aligned along the direction of the (7×7) unit cell, along $[01\bar{1}]$. Time-resolved images of the $(7 \times 7) \Rightarrow (1 \times 1)$ transition obtained at constant temperature, acquired approximately every 17 s (Miki et al., 1992) in this range of temperatures, indicate nucleation of the (7×7) from step edges with fluctuations in the domain boundaries reduced as the temperature is lowered. The $[\bar{1}\bar{1}2]$ steps straighten as the (7×7) domain forms, because the unfaulted half of the unit cell is lower in energy that the faulted half. This leads to an equilateral triangular shape to the (7×7) domains bounded by unfaulted triangles.

On vicinal Si(111) surfaces misoriented to $[11\bar{2}]$, slender (111) facets containing a few (7×7) unit cells nucleate; these widen in quanta of the (7×7) unit cell width as the temperature is lowered (Hibino et al.,

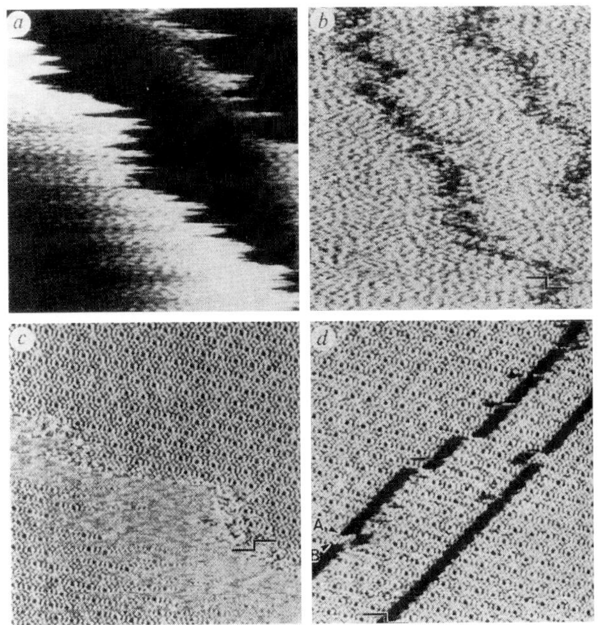

Figure 41. STM images of an annealed Si(111) surface showing the formation of the (7 × 7) from the high-temperature (1 × 1) structure. (a) 100 nm × 100 nm constant current image acquired at 880 °C; (b)–(d) current images acquired at 880 °C (50 nm × 50 nm), 860 °C (100 nm × 100 nm), and 840 °C (100 nm × 100 nm), respectively (Kitamura et al., 1991).

1993). These are separated by step bunches, which below 700 °C are transformed into (12 × 1) reconstructed (331) facets that coexist with the (111) facets which now contain both (5 × 5) and (7 × 7) structures. The (5 × 5) is more stable under compressive stress which might occur due to the presence of the steps.

2.4.2 Si(100)

Si(100)-2 × 1

The Si(100) surface is the technologically-important Si substrate material for microelectronics device fabrication. On a flat surface, locally-ordered 2 × 1 domains exist separated by monoatomic steps which produce a 90° rotation between them. Thus diffraction techniques such as LEED average over both domains. In addition, averaging techniques are less sensitive to the lack of atomic-scale structural homogeneity that may exist to a large degree, depending on surface preparation. Vicinal surfaces provide a means to obtain a dominance of one orientation and, varying the miscut angle and off-axis direction, is a means to study terrace-step structures (Wierenga et al., 1987) and analyze kink-step energetics (Swartzentruber et al., 1990).

In the late 1970s, theoretical calculations and experimental observations of the electronic structure deduced that the reconstruction consisted of an ordered arrangement of Si surface dimers, which form to reduce the number of dangling bonds (Chadi, 1979). A further lowering of energy was predicted if an ordered (tilted) array of asymmetric dimers form since symmetric dimers should be unstable with respect to a Peierls distortion. This would render the surface non-metallic, as observed by STS (Hamers et al., 1986 b) and other spectroscopies.

The first STM images of this surface (Hamers et al., 1986 b) directly confirmed that dimers are the basic building blocks of the reconstruction, as seen in Fig. 42. Bias-dependent images showed that the occu-

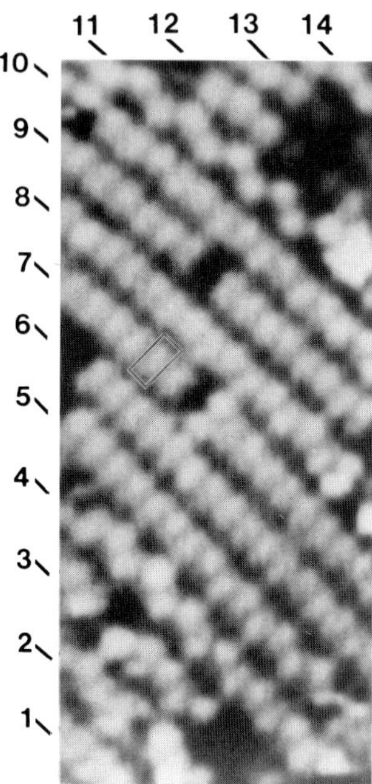

Figure 42. Occupied state topographic image of Si(001)-2 × 1 (Hamers et al., 1986 b).

pied state density is between the dimer atoms, while the unoccupied states are localized away from the dimer, as seen in Fig. 22. STM indicates that, in the absence of defects, symmetric dimers compose the terrace structures; near defects, however, a clear asymmetry is apparent. In Fig. 42, row 11 is a defect-free dimer row; several missing-dimer defects exist in other rows (e.g., row 9). Sometimes around such defects, the dimer rows appear as zigzag structures forming superstructures, which suggests that a correlated tilting of the dimers occurs due to dimer–dimer interactions.

At low temperature, LEED has provided evidence of higher-order structures with an order ⇒ disorder transition occurring at 220 K (Tabata et al., 1987). Ruling out tip-induced surface distortions (Huang and Allen, 1992), the negative observations of asymmetry in defect-free regions with room temperature STM led to the speculation that the dimers are oscillating between two tilted configurations and that, on the timescale of the STM measurement, these average to form a symmetric image (Hamers et al., 1986 b). Molecular dynamics simulations supported the picture of dimer oscillations within the time scale of the STM measurement (Gryko and Allen, 1992). At 300 K, the dimers switch between left-asymmetric / symmetric / right-asymmetric configurations and this switching is quenched at 200 K. A significant low temperature (120 K) STM study of the Si(100) surface confirmed this model (Wolkow, 1992). The image in Fig. 43 shows that an increased number of buckled-dimer domains occur on defect-free regions at 120 K. Some defects pin dimers into buckled configurations to inhibit good ordering due to strain effects. Explaining the low temperature observations requires adopting a double-well asymmetric potential, which further explains the non-metallic nature of this surface.

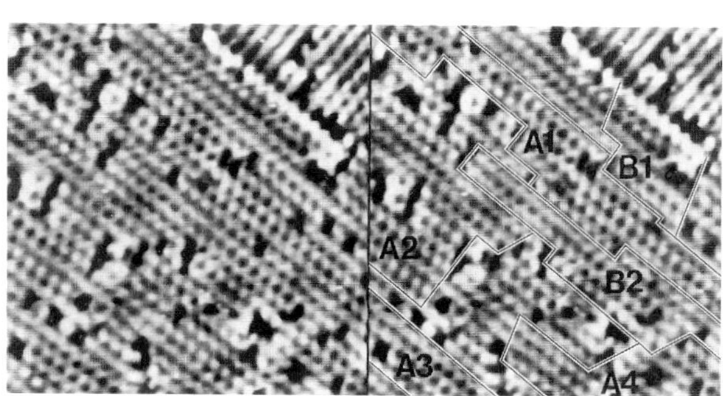

Figure 43. 200 Å × 200 Å image of Si(001) acquired at low temperature (120 K) with $V_t = 2$ V. Domain boundaries are marked in the right panel (Wolkow, 1992).

Initial Stages of Epitaxial Growth

Molecular beam epitaxy is an established method for obtaining atomically-perfect semiconductor interfaces. Observing the initial stages of growth at the atomic level under various processing conditions provides a means to directly determine nucleation and growth processes. Homoepitaxial growth of Si on Si(001) was one of the first examples in which the growth kinetics were related to the ultimate equilibrium structures (Mo et al., 1989). Si deposition to sub-monolayer coverages was performed at sample temperatures between 300 K and 500 K. In this regime, Si on Si(001) exhibits anisotropic islands, as shown in Fig. 44. Ordering for such a system containing two coexisting phases (islands and 2D vapor) involves nucleation, growth, and coarsening, and is ultimately controlled by differences in boundary free energies between islands of various sizes and shapes. An immediate possibility to explain the observed anisotropic island shapes is to assume an anisotropic diffusion coefficient. This is likely because the substrate surface, consisting of dimer rows, is itself anisotropic. Another possibility is an anisotropic bond energy to the island boundaries. Modeling anisotropic diffusion coefficients, however, led to a negative prediction for island shape anisotropy; this is consistent with the additional observation that annealing made all the islands more isotropic with an equilibrium shape anisotropy of ~ 3. Modeling the accommodation coefficient at the side of a growing dimer chain to be considerably less (~ 0.1) than at the chain end led to the observed island shape anisotropy.

Further questions arise as to the diffusion coefficient of Si on Si(001) and the effect of steps on the growth process. Regarding the latter, it was observed that the atoms do not attach at steps to form a single-height step but immediately form double-height steps (Mo et al., 1990b). This effect implies that the diffusion coefficient across steps is rather high. The diffusion coefficient, D, was measured (Mo, 1991) by statistically analyzing the number density of islands during deposition within a model relating D to island formation. If D is large, there is a greater chance for the growth of exsting islands, while if D is small, there is a higher probability to nu-

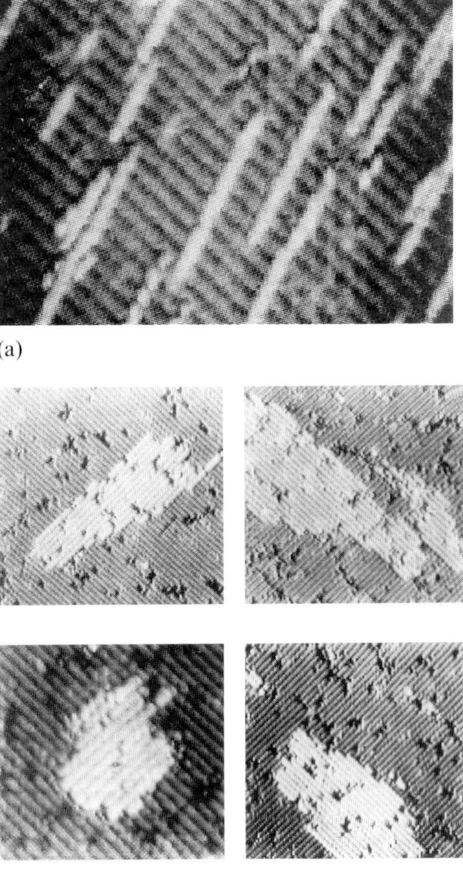

(a)

(b)

Figure 44. (a) 230 Å × 230 Å topographic image of 0.005 ML Si deposited at 475 K on Si(001). (b) Island shapes become more rounded with less anisotropy after annealing at ~ 600 K for ~ 120 s (Mo et al., 1989).

cleate new islands. Using anisotropic bonding as before and a diffusion anisotropy of 1000 (Mo, 1991), the value of D was found to be 0.67 eV with a pre-factor of $\sim 10^{-3}$ cm^2/s.

Heteroepitaxial growth was studied in STM measurements of Ge on Si(001) (Mo et al., 1990). Ge growth on Si(001) follows the Stranski–Krastanov process in which the deposited material wets the surface forming a strained layer [$\sim$ 3 atomic layers for Ge/Si(001)]; when the surface energy of the strained layer becomes sufficiently low, three-dimensional (3D) clusters begin to form. It was found that the kinetics of 3D cluster formation proceeds via small intermediate (105) Ge facets that form on the strained layer. These clusters promote the accommodation of arriving atoms better than direct nucleation of new 3D clusters.

2.4.3 GaAs(110)

Understanding the surface properties of III–V compound semiconductor surfaces and interfaces is important for new device applications. Since the surface and interface properties of such compounds as GaAs are typically rather different from Si, STM studies of these materials have led to novel results directly related to the nature of the surfaces of compound semiconductors. In Sect. 1, it was noted that voltage-dependent imaging of GaAs(110) is the prototype system to demonstrate the ability of STM to directly relate geometric and electronic structure. Chemical identification using valence band structure is typically difficult, but surface chemical analysis is surprisingly straightforward since the anion and cation species can be differentiated by comparing images acquired with opposite polarities (Feenstra et al., 1987b). In Fig. 4, boxes locate identical regions for which images with $V_t = -1.9$ V probe unoccupied states and $V_t = +1.9$ V probe occupied states. It is immediately apparent that (1) the images are displaced from each other both along $[00\bar{1}]$ and along $[1\bar{1}0]$ and (2) the shapes of individual features for opposite polarities are different. The localization of electrons (occupied states) about the As atoms and the localization of an empty orbital about the Ga atoms explains the lateral difference in the images. The separation in energy between the surface states reflects changes in the surface structure caused by the surface electronic structure. The detailed electronic structure was used to analyze the clean surface lattice relaxation. It was confirmed that the Ga–As rows are rotated by $\sim 28°$ driving the surface bond hybridization towards a (planar) sp^2 configuration to lower the surface energy.

2.4.4 Photoinduced Processes

Semiconductor substrates are usually rather highly doped so as to provide sufficient room temperature conductivity for STM imaging without inducing a voltage drop across the substrate. Inducing thermally-stimulated carriers by heating the sample has been demonstrated as one means to increase the number of carriers for tunneling (Binnig, 1985). Illuminating the sample to form a photoconductive top layer was demonstrated as an alternative method to permit imaging of samples of low conductivity (van de Walle et al., 1987). In this study, an HeNe laser and a halogen lamp permitted imaging of a semi-insulating ($\varrho \approx 7 \times 10^7$ Ω cm) GaAs wafer and was presented as a means to image other insulators with STM.

Optical illumination in combination with STM has also been used to study excited carrier relaxation on the nanoscale (Hamers and Cahill, 1990; Hamers and

Markert, 1990). Under DC illumination, it is possible to measure a voltage created across the space-charge region by optically-created electron–hole pairs which form a non-equilibrium charge distribution. In one study on an Si(111)-(7 × 7) surface, this local surface photovoltage (SPV) was measured as a function of illumination intensity and position with the STM operating in a potentiometry mode (Hamers and Markert, 1990). The images showed that defects decrease the SPV in a region ~ 25 Å in diameter. In these regions, the model developed suggested a recombination rate approximately four times higher in the defect region than in the ordered (7 × 7) surface. A time-resolved study of carrier lifetimes on an Si(111)-(7 × 7) surface was performed using a picosecond laser pump and the STM as a synchronous probe operated in a scanning capacitance arrangement (Hamers and Cahill, 1990). The time dependence of the SPV, as shown in Fig. 45 was fit in terms of a simple model, where the SPV decay is $\tau = \tau_D \cdot \exp\{-e[SPV(t)]/(kT)\}$, the value of τ_D deduced from the data was ~1 μs.

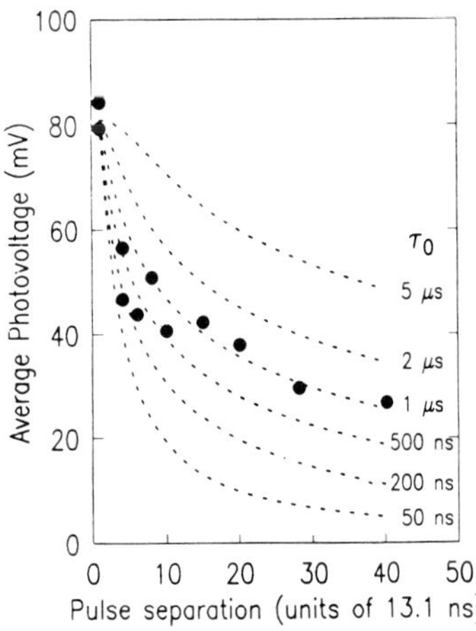

Figure 45. Time-dependence of surface photovoltage as a function of delay time between optical pulses (Hamers and Cahill, 1990).

2.5 Metal–Semiconductor Interfaces

Several aspects of advanced microelectronics device technology require an understanding of the structure and electronic properties of metal–semiconductor contacts in atomistic detail. This assertion is justified by the demonstration that the Schottky-barrier heights are intimately related to the structural details at abrupt epitaxial interfaces (Tung, 1984; Heslinga et al., 1990); these observations differ markedly from the classical theory of Schottky-barrier formation (Schottky, 1938; Schottky, 1940). Ulimately controlling metal film growth requires the control of chemistry, i.e. avoiding undesirable compound formation, and growth kinetics, to achieve epitaxial structures.

Development of the macroscopic metal–semiconductor interface, with its corresponding electronic properties, begins with sub-monolayer coverages of metal atoms, passes through the process of metallization, and continues to the development of a multilayer film, which may exhibit a rectifying or an ohmic behavior. Metal films typically do not grow epitaxially layer-by-layer (Frank–van der Merwe) on semiconductor surfaces due to the surface kinetics at the substrate temperatures at which deposition is typically performed. Two alternate growth modes are Stranski–Krastanov, involving the initial formation of a monolayer followed by 3D island (cluster) growth, or Volmer–Weber, involving only 3D island growth. STM can be used to measure the morphology of the surface in multilayer regime and island sizes and their distributions directly under various processing conditions. Metallic islands formed on a semiconducting surface are expected to undergo an insulator → metal transition at some critical three-dimensional size, and the transition can be characterized by STS. To assess metallic behavior by detection of electronic states at E_F, i.e. under zero-bias conditions, the spectroscopic dynamic range can be enhanced by taking current images at several sample–tip separations (Feenstra, 1989) or by advancing the probe tip towards the sample during a voltage sweep (Feenstra and Mårtensson, 1988).

One fundamental aspect of this problem, with roots in surface science, involves the evolution of the metal–semiconductor interface in its initial stages. Thin film science takes over in the sub-monolayer to few-layer regime, where imaging provides input into growth processes, and spectroscopy can elucidate the chemical bonding interactions as well as the evolution of electronic properties. Finally, when the macroscopic thin film is complete, the relation between the electronic properties of the initial interface and that of the final metal–semiconductor junction may depend strongly on the unique properties of the buried interface. Cases in which STM has been employed to study metal–semiconductor interfaces in these regimes are discussed below.

2.5.1 Alkali-Metal–Semiconductor Interfaces

Alkali-metal–semiconductor interfaces have been studied extensively as model metal–semiconductor systems, because the electronic structure of an alkali-metal atom is simple, chemical reactivity with the substrate is typically small, and a variety of ordered structures may exist in the first monolayer of growth. For example, STM studies indicated that Cs grows in one-dimensional chain structures on III–V(110) surfaces for very low coverages (< 0.2 ML). This suggests the fascinating possibility of atom-thick lines, or "atomic wires". STS, however, found these structures to be insulating, but close-packed structures at higher coverages appeared to approach metallic behavior. In order to understand these observations, the dimensionality of the structure and the local density must be considered.

Figure 46 A shows that the 1D structures at ~0.2 ML consist of well-separated atomic Cs rows arranged in zigzag structures on GaAs(110) (Whitman et al., 1991 a, c); similar structures were observed on InSb(110) (Whitman et al., 1991 c). The classical picture of electron transfer between alkali-metal atoms and the substrate would predict ionic bonding and repulsive adsorbate–adsorbate interactions. However, this scenario is unlikely because the STM image indicates that anisotropic attractive interactions between Cs atoms induce the formation of Cs rows at the lowest coverages. As the coverage increases, triple chains that form local $c(2\times2)$ structures, and the room temperature saturated monolayer results in the formation of local $c(4\times4)$ structures. Although poor long-range order causes diffraction measurements to be insensitive to local structure, STM revealed that the $c(4\times4)$ structures are 5-atom Cs crystallites that correspond to the surface structure of Cs(110). Electron counting dictates that these structures should be metallic in a single-particle picture. In fact, at room-temperature saturation coverage, the packing density approaches that of metallic Cs. As shown in Fig. 46 B, the zero-bias conductivity indicates that the lower coverage structures are actually non-metallic, although the gap has narrowed. These results are consistent with recent studies of the spatially-averaged electronic excitation spectra under similar conditions, which show that the Cs/GaAs(110) interface is non-metallic due to many-body effects in the ultrathin

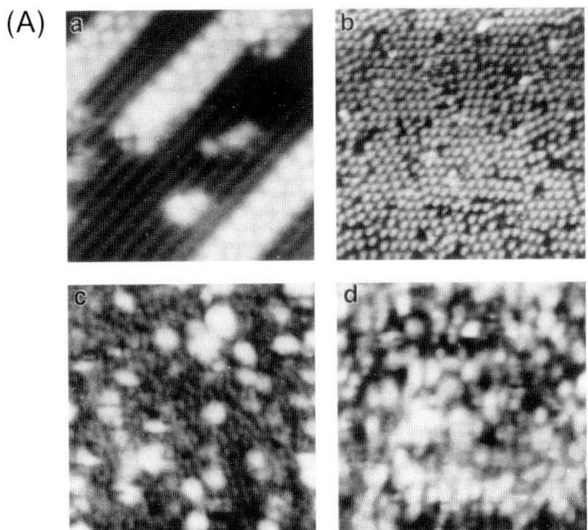

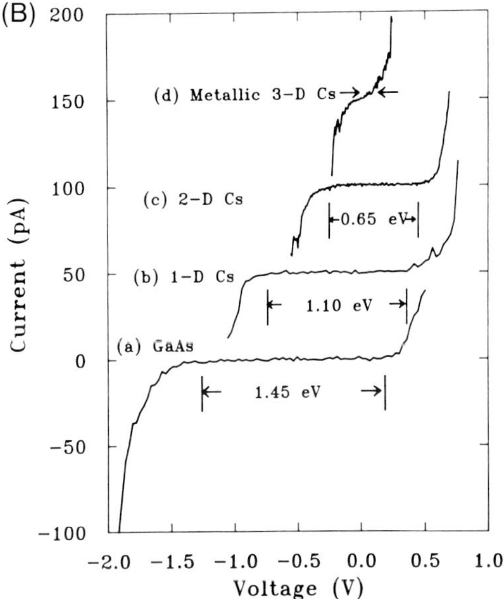

Figure 46. (A) STM image of one-dimensional Cs structure and of the $c(2\times2)$ structure at higher Cs coverage on GaAs(110). (B) STS spectra for these regimes (Whitman et al., 1991 a).

overlayer. Further spectroscopic measurements on this system at room temperature indicated metallization when (three-dimensional) bilayers formed at higher coverages. On the basis of electron energy loss measurements there is some disagreement as to whether a gap still exists so that these structures would be non-metallic at zero temperature (DiNardo et al., 1990; Weitering et al., 1993 a).

2.5.2 Growth of Trivalent Metals on Si(001)

LEED and RHEED (reflective high-energy electron diffraction) studies have found that Al, Ga, and In form similar coverage-dependent series of ordered structures on the Si(001)2×1 surface (Knall et al., 1986; Bourgiugnon et al., 1988 a, b; Ide et al., 1989). Subsequent STM studies have provided details of these structures in the sub-monolayer regime (Baski et al., 1990; Baski et al., 1991; Nogami et al., 1991). Taking Al/Si(001) as an example, around room temperature an epitaxial 2×2 structure forms at 0.5 ML; this is preceded by the growth of one-dimensional Al structures (Ide et al., 1989). The local structure comprising these overlayers represents Al ad-dimers which form in a direction orthogonal to the clean Si(001)-2×1 surface dimers (Nogami et al., 1991). Occupied state STM images shown in Fig. 47 illustrate the evolution of these ad-dimer chains from 0.1 ML to 0.5 ML. Bias-dependent images at 0.4 ML are also shown in Fig. 47. Unfortunately, since individual Al atoms could not be identified, due in large part to the complexity of the valence structure, the sites of Al on the Si surface were not unambiguously resolved. Beyond 0.5 ML, the first layer is saturated and Al clusters begin to form in a second layer in a Stranski–Krastanov growth mode.

Recent calculations illustrate the mechanism for symmetric ad-dimer formation (Brocks et al., 1993) and account for the bias-dependent images observed. Single Al atoms, which are not observed on the surface, are free to diffuse in any direction at room temperature with a small energy

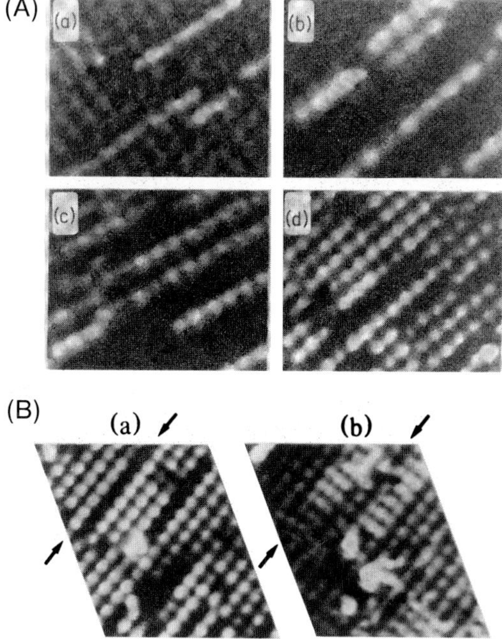

Figure 47. (A) STM images showing the evolution of the Al/Si(001) interface at submonolayer coverages: (a) 0.1 ML, (b) 0.2 ML, (c) 0.3 ML, (d) 0.4 ML. (B) Images of 0.4 ML Al-covered surface acquired with (a) $V_t = -2$ V and (b) $V_t = +2$ V (Nogami et al., 1991).

minimum at the symmetric hollow site between two pairs of Si dimers. Adding a second Al atom on the surface results in the formation of a parallel ad-dimer between the substrate surface dimers without disruption, as shown in Fig. 47 and in agreement with previous calculations (Northrup et al., 1991). This configuration is preferred over ad-dimer configurations, either between the Si dimers and rotated by 90°, or forced to reside in the Si dimer rows, because favorable bond angles with minimal strain are possible. This provides a natural mechanism for the formation of the observed dimer chains oriented perpendicular to the clean surface dimer rows. A third adatom preferentially binds at a reactive site located next to the existing Al dimer and situated energetically near E_F. A fourth Al atom achieves a minimum energy configuration by forming a second dimer. This process may be viewed as a surface polymerization reaction (Brocks et al., 1993). Electronically, the ad-dimer interacts with the π and π^* states of the clean Si(001) dimers. The occupied local density of states near E_F is localized near the Si atoms that have reacted with Al, as shown in Fig. 47. This represents remnants of the disrupted, but not broken, π bond of the clean surface. In the empty state image, the unoccupied dangling bond states of Al are expected to appear between the Si dimers. This model is in qualitative agreement with experimental observations (Nogami et al., 1991). It is also expected that, as the coverage is increased, repulsive interactions will create dimer rows with $p(2 \times n)$ periodicity.

2.5.3 Ambiguities in Structural Determinations

The exclusive use of STM as an imaging or even as a spectroscopic probe in order to obtain direct structural information must be done with care, especially in metal–semiconductor systems. There are examples in the literature where controversy has developed around the interpretation of STM images to obtain structural models for metal–semiconductor systems. Since the valence structure which is probed is likely to change dramatically as the metal–semiconductor interface develops good insight into the evolution of local electronic structure from theoretical calculations as well as element-specific probes [e.g., Auger, X-ray photoelectron spectroscopy (XPS), Rutherford backscattering (RBS)] to provide absolute and relative coverages, and structural probes extending beneath the top layer, are crucial to couple with STM data. Recently, sophisticated theoretical calculations have made great progress in testing various structural models to provide simulated STM images.

We briefly consider the Ag/Si(111)-$\sqrt{3} \times \sqrt{3}$ and the Na/Si(111)-3×1 systems as examples where complementary techniques and high-quality theory have been required to provide a fully consistent interpretation of the STM images to arrive at the structure.

Si(111)-$\sqrt{3} \times \sqrt{3}$ Ag

Obtaining the structure of the Si(111)-$\sqrt{3} \times \sqrt{3}$ Ag interface has been particularly difficult over the years because of seemingly conflicting results from an array of surface science techniques. A few of the structures that have been proposed are shown in Table 1. Here a knowledge of the absolute Ag coverage can be used to distinguish between structural models.

STM studies of the initial stages of Ag nucleation on the Si(111)7×7 surface at $\leq 130\,°C$ shows that Ag atoms bond to the dangling bonds of the second atomic layer in the 7×7 unit cell with the subsequent formation of two-dimensional islands of close-packed Ag atoms while the 7×7 unit cell is preserved (Tosch and Neddermeyer, 1988). A 3×1 structure can be formed by exposure to an Ag flux above $\sim 500\,°C$ (Wilson and Chiang, 1987b) (see below). The Si(111)-$\sqrt{3} \times \sqrt{3}$ Ag structure forms after deposition of ~ 1 ML Ag at $\sim 460\,°C$ (van Loenen et al., 1987), or for room temperature (RT) deposition followed by annealing at 200–500 °C (Wilson and Chiang, 1987b).

Table 1. Selected proposed Si(111)-$\sqrt{3} \times \sqrt{3}$ Ag structures and Ag coverages.

Structure	Ag coverage
Embedded trimer (ET)	1 ML
Honeycomb (H)	2/3 ML
Honeycomb-chain-trimer (HCT)	2/3 ML

Occupied- and unoccupied-state images of the Si(111)-$\sqrt{3} \times \sqrt{3}$ Ag structure are shown in Fig. 48 (Wan et al., 1992b). An early STM study of structure (van Loenen et al., 1987) decided on the ET model based on CITS spectra, the non-metallic nature of the surface, and prior ion scattering results. This picture requires that the imaged atoms are Si atoms atop the embedded Ag atoms. Another set of studies decided upon the H model, assuming that Ag atoms form the image (Wilson and Chiang, 1987b), and by further considering the registration of the $\sqrt{3}$ structure with respect to the 7×7 nucleated at lower Ag coverages (Wilson and Chiang, 1987a). Two theoretical calculations, however, found that the HCT model produces a consistent explanation for the majority of surface science studies, including further bias-dependent STM studies. In the latter, the registration arguments previously put forth were shown not to be unique depending on the presence or absence of a step at the domain boundary. While this structure finally appears to be resolved on the basis of comparisons of STM (Wan et al., 1992b) and other techniques, it is clear that calculations of total energy values for several structures (Ding et al., 1991), along with comparisons of the calculated electronic structure (Ding et al., 1991; Watanabe et al., 1991) with bias-dependent imaging, are essential in order to perform a dependable structural analysis on the basis of the spatial-dependence of the valence electronic structure.

3×1

The Si(111) surface exhibits an adsorbate-induced 3×1 reconstruction when it

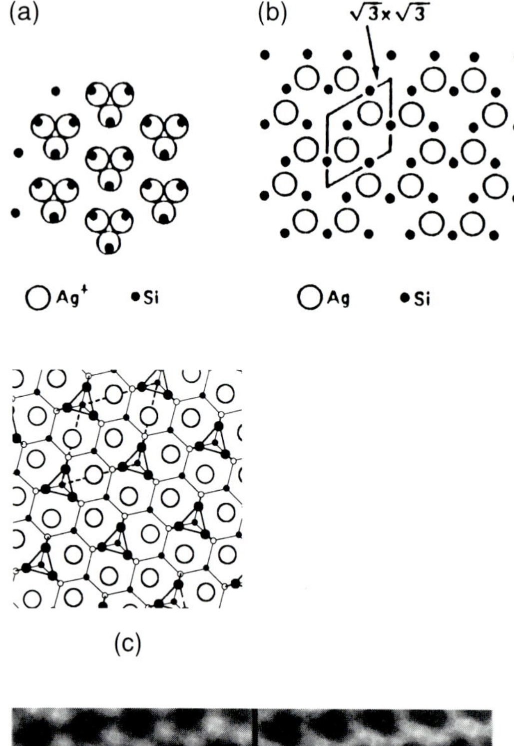

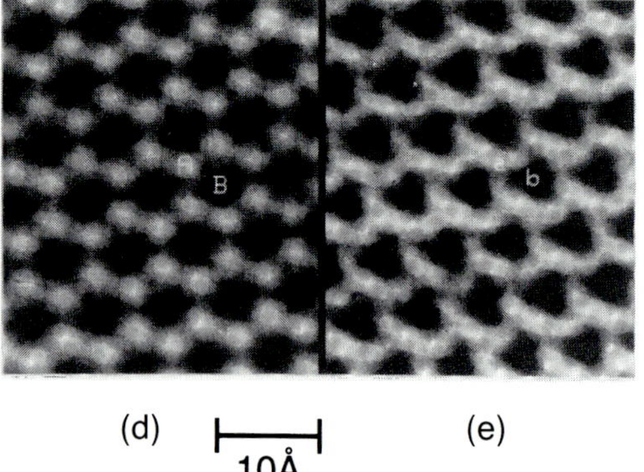

Figure 48. (a)–(c) Proposed models for the Si(111) ($\sqrt{3} \times \sqrt{3}$)-Ag interface – embedded trimer, honeycomb, and honeycomb-chain-trimer. STM images of (d) empty states and (e) filled states for the Si(111) ($\sqrt{3} \times \sqrt{3}$)-Ag interface (Wan et al., 1992b).

is exposed to Ag, Li, Na, K, or H at moderately high temperatures (300–600 °C). It is generally agreed that a sub-monolayer of the adsorbate remains when forming the 3×1 reconstruction. Making the assumption that STM images comprise the metal atom and without complementary techniques, the Na-, Li-, or Ag-induced 3×1 have been interpreted in terms of an ordered reconstruction with an overlayer coverage of 2/3. It is agreed that a channel along ⟨110⟩ separates the rows of bright structure forming the 3×1, as shown in Fig. 49.

Consideration of other data, however, argues against a 2/3 coverage and suggests that rehybridized Si atoms form the surface image with a 1/3 coverage (Weitering et al., 1993; Weitering and Perez, 1993). LEED studies have obtained similar $I-V$ spectra for all of these systems. This result suggests that the adsorbate is present in very small quantities, disordered, and/or far enough beneath the surface such that there is minimal contribution to the $I-V$ spectra. Comparisons of Auger spectral intensities for Na (K) in the 3×1 surface versus Na (K) adsorbed on Si(111)-7×7 at room temperature indicate ≤ 0.33 ML Na (K) coverage. It is clear that the resolution of this structure will emerge only when coverage and electronic structure information is consistent with the model of geometric structure.

2.5.4 Electron Localization at Defects in Epitaxial Layers

The transition from non-metallic to metallic behavior in an ultrathin metal film grown on a semiconductor surface involves the evolution from localized to delocalized electronic states. In some cases, the non-metallic state of an ultrathin metal film cannot be explained by a single-particle picture and a many-electron description must be used. The electronic states of isolated defect structures possess a high degree of spatial localization where coulomb interactions and electron correlation effects require such a description. The imaging and spectroscopic capabilities of STM

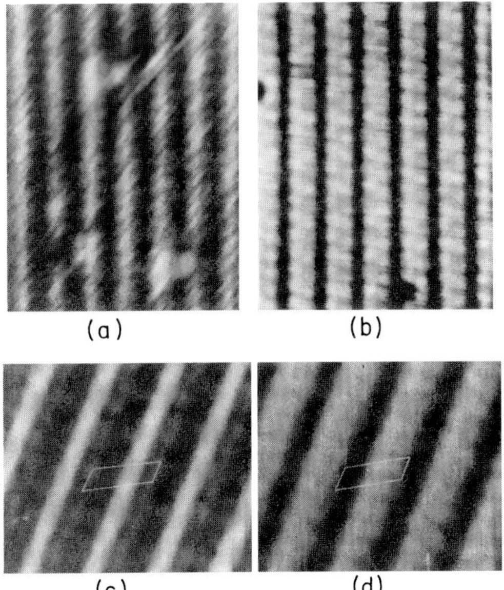

Figure 49. Occupied (a, c) and unoccupied (b, d) state STM images of the Ag-induced Si(111)-3 × 1 structure showing the "missing-row" reconstruction (Wan et al., 1992 a).

were used to demonstrate this phenomenon for the dangling-bond defect of the Al/Si(111) interface. The image of the Si(111)-Al($\sqrt{3} \times \sqrt{3}$) interface, shown in Fig. 50 a, contains several defects labeled "S" that appear dark, with $V_t = -2$ V, and appear to be enhanced with respect to the normal $\sqrt{3} \times \sqrt{3}$ Al unit cells (labeled "I"), with $V_t = 2$ V. The number of S-type defects decreases as the Al coverage approaches 1/3; this suggests that these S- defects are Si adatoms which replace the Al adatoms of the ideal interface. Figure 50 b shows that the energy spectrum measured above an S-type defect exhibits an occupied state, at ($E_F - 0.4$ eV), with significant density between ($E_F - 1$ eV) and E_F. There is a minimum state density at E_F and a new unoccupied state at (E_F + 0.9 eV). Previous photoemission studies attributed the occupied state at 0.4 eV to domain boundaries, but local imaging/spectroscopy clearly demonstrates that isolated defects are the origin of this feature. In contrast to an Al adatom, an Si adatom has a dangling bond that extends into the vacuum with a negligible energy dispersion. Besides the absence of states at E_F, STS shows there is little interaction with the surrounding surface and no measurable charging that would result in local band bending. In a single-particle picture, the half-filled dangling bond of these defects would produce a state at E_F. However, in the theory of localized defects, three charge states exist. The neutral state ($n = 1$) exhibits a bimodal state distribution with donor and acceptor levels below and above E_F, respectively. Measurement of the occupied state corresponds to an $n = 1 \rightarrow n = 0$ transition probing the donor level; the unoccupied state measurement is an $n = 1 \rightarrow n = 2$ transition

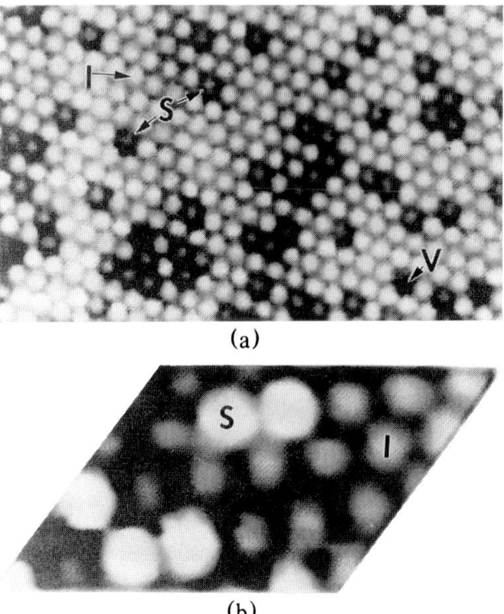

Figure 50. STM images of Si(111) ($\sqrt{3} \times \sqrt{3}$)-Al at (a) $V_t = -2$ V and (b) $V_t = 2$ V showing S-type defects and I-type defects. The S-type defects are dark with a bright center in the unoccupied state image and are bright in the occupied state image (Hamers and Demuth, 1988).

which probes the acceptor level. According to STS, the Si adatom has a donor level at ($E_F - 0.4$ eV) while the acceptor level is at or above the CBM. By this analysis, the Coulomb energy due to electron localization is $U_{eff} \geq 0.4$ eV, which is in agreement with theoretical predictions.

2.5.5 E_F Pinning, Mid-Gap States, and Metallization

In the initial stages of growth, thin metal-atom layers typically exhibit a band gap before metallization. Focusing on the electronic properties of the abrupt interface itself, we see that, at lowest metal-atom coverages, new electronic states derived from the bond of the metal-atoms to specific surface sites may exist in the substrate gap resulting in shifts of the valence band maximum (VBM) position with respect to E_F, eventually pinning the Fermi level position. Realistically, defects, isolated atoms, and island or cluster edges are likely to introduce new states in the gap possibly resulting in a different E_F position than if the ultrathin film exhibited ideal epitaxy. Evolution of the local electronic structure in the initial stages of thin film growth was first obtained through comparisons of Au, Sb, and Bi on GaAs(110) using STM/STS. This information is depicted in Figs. 51, 52 and 53.

Following the growth of the metal–semiconductor interface from isolated atoms provides a basis for a local origin to gap states which pin the Fermi level. The

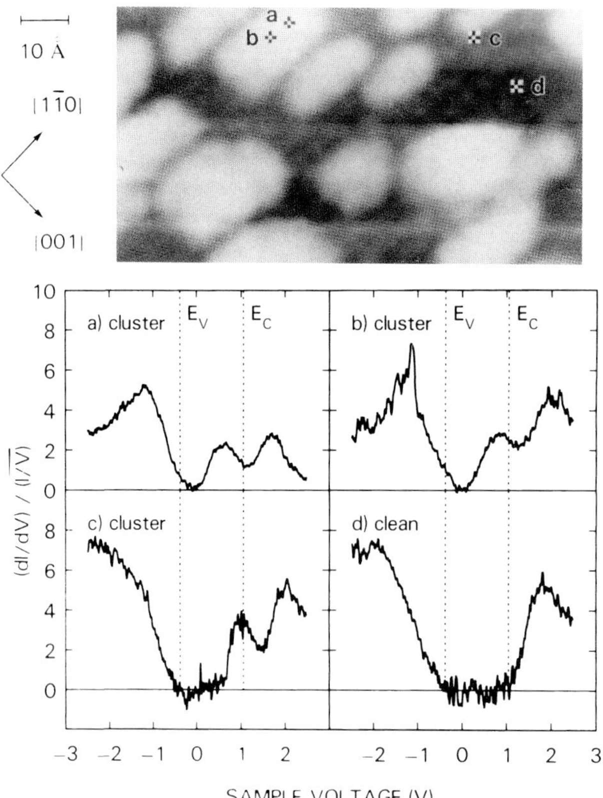

Figure 51. 100 Å × 55 Å occupied state image of 0.25 ML Au on GaAs(110) and associated electronic spectra (Feenstra, 1989).

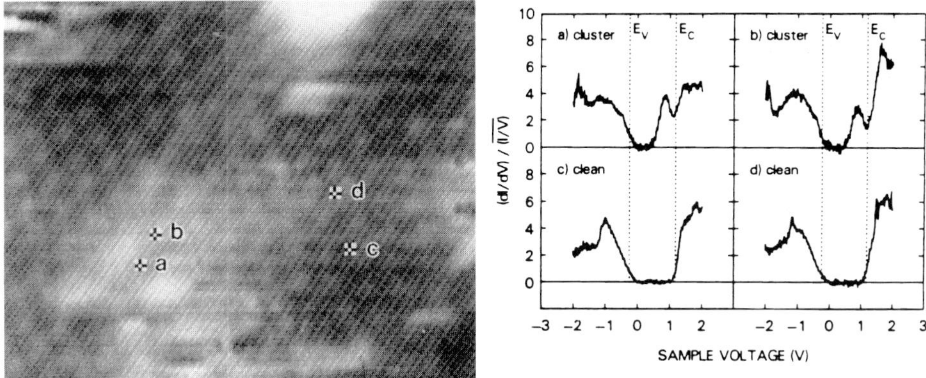

Figure 52. STM image and spectra on top of and on the edge of an Sb island on GaAs(110) (Feenstra et al., 1989a).

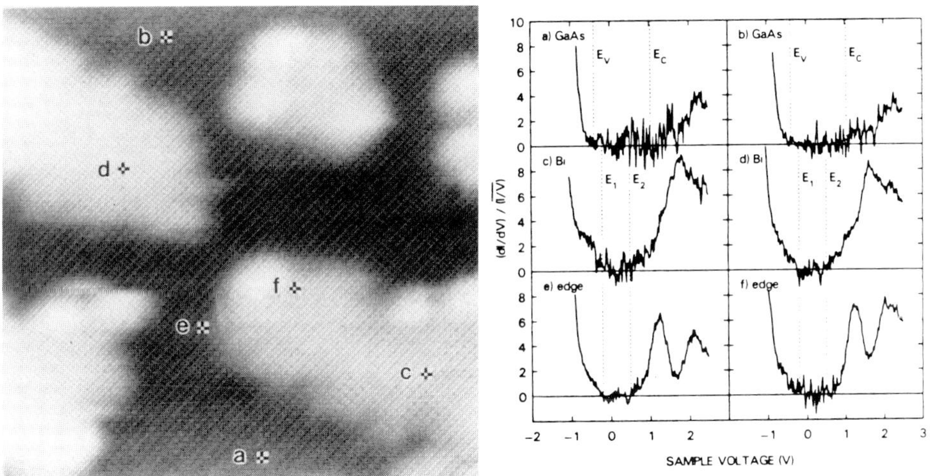

Figure 53. STM image and spectra on and off of Bi terraces on GaAs(110) (Feenstra et al., 1989a).

deposition of Au onto GaAs(110) follows a Vollmer–Weber growth mode. Au atoms bond to Ga sites and this covalent bond is a mixture of the Au 6s and (empty) Ga sp^3 orbitals (Feenstra, 1989). STS spectra above isolated Au atoms or few-atom structures showed that the bonding orbital sits below the VBM at ($E_{VBM}-0.7$ eV) and a gap state exists at ($E_{VBM}+1.0$ eV). For these isolated structures, the spatially-coincident filled resonance state and empty gap state are viewed as donor and acceptor levels, respectively. Increased coverages of Au produce clusters displaying Au(111) facets. Following the evolution of spectra as the film thickness increases, the gap states are found to remain intact (although they broaden) until the contiguous metal layer masks these states in favor of the metallic continuum.

Sb deposition allows growth to commence and form an ordered Sb overlayer (0.5 ML) that consists of zigzag Sb chains bridging chains of Ga and As surface atoms. At higher coverages, Sb clusters show up as protrusions, or defects, upon some of the terraces. STS on Sb terraces show no mid-gap states but an observed spectral shift provides evidence for band-bending of ~ 0.2 eV. Sb islands produce additional band bending ~ 0.4 eV that is reduced by screening as the tip is moved away from the island. The study of > 0.5 ML disordered Sb films provided evidence that the Fermi level is pinned between two narrow electronic states located at the edges of Sb islands associated with unsatisfied surface bonds.

Bi/GaAs(110) produces similar 1×1 terrace structures as Sb, but, due to the larger metallic radius of Bi, misfit dislocations occur about every six rows. For ordered Bi terraces, the Bi states extend into the gap resulting in a narrowing of the band gap, and E_F is pinned at the surface VBM. For Bi terrace edges, however, a new unoccupied state appears above the band gap edge which, therefore, does not modify the E_F position.

2.5.6 The Insulator → Metal Transition in Metallic Fe Clusters Grown on Semiconductor Surfaces

Fe grown on GaAs(110) provides another example of Vollmer–Weber growth. For this system, quantum size effects prior to metallization have been observed with STS (Dragoset et al., 1989; First et al., 1989 b). Figure 54 shows an STM image of 0.1 Å Fe on *p*-type GaAs(110) where several clusters have formed. Comparisons of spectra from a "thick" (17 Å) Fe film, a 1150 $Å^3$ cluster, a 150 $Å^3$ cluster, and clean GaAs(110) show that clusters up to ~ 400 $Å^3$ (~ 35 atoms) are non-metallic, i.e. a gap of 0.1–0.5 eV exists, while clusters of greater size exhibit $I-V$ characteristics similar to the bulk metal. The quantum

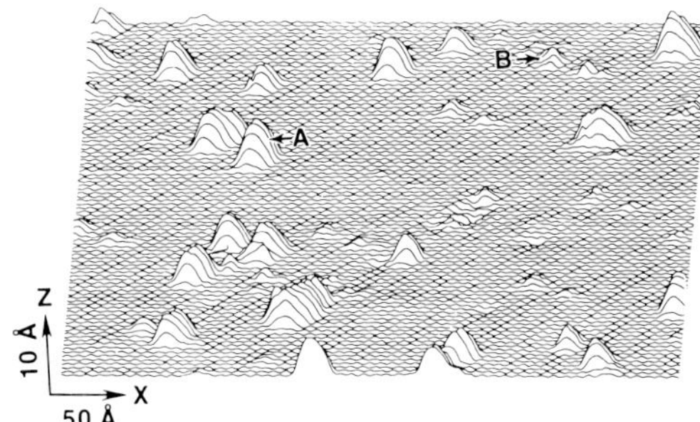

Figure 54. 400 Å × 384 Å image of 0.1 Å Fe on GaAs(110) with $V_t = 2.5$ V (First et al., 1989 b).

size effects that result in a band gap occur when the spacing between energy levels is of the order of a few $k_B T$, which is consistent with metallic behavior being observed at room temperature for ~ 30 atoms. In addition, near metallic Fe clusters, evanescent gap states, which have been associated with the theory of metal-semiconductor interfaces through the years (Tersoff, 1984), which decay exponentially with distance away from a cluster, were observed.

2.5.7 Microscopy and Spectroscopy of Buried Interfaces – BEEM

This rectifying nature or metal–semiconductor interfaces is based on the classic Mott–Schottky model. In this mode, the Schottky-barrier height (SBH) can be related to the macroscopic properties of the semiconductor and the metal, i.e., SBH $= \Phi - \chi$, where Φ is the work function of the metal and χ is the electron affinity of the semiconductor. It is widely recognized that the actual situation is considerably more complicated. First, the interface between a metal film and a semiconductor surface is often inhomogeneous. In addition, chemical reactions or interdiffusion may occur at an interface, further altering its identity. Even with high-quality abrupt epitaxial interfaces, differences in the atomic structure at the atomically-abrupt interface can greatly affect the SBH (Tung, 1984; Heslinga et al., 1990). Spectroscopic pictures of the first layer(s) of metal atoms, as described in the previous section, must be extended to obtain a view underneath multiple metal layers in order to explain the local properties of a macroscopic metal–semiconductor interface. Electron spectroscopies are typically not sensitive of the subsurface region; the small mean free path of electrons causes the surface to dominate. Optical techniques hold promise but are not position sensitive on the nanoscale. The fact that hot electrons can travel without energy loss – the ballistic transport regime – through thin metal films makes ballistic electron emission microscopy (BEEM) (Kaiser and Bell, 1988) an ideal method to measure local SBHs and band offsets, and to image the local structure of buried interfaces.

Some of the theoretical aspects of BEEM have already been discussed. In order to measure the SBH between a metal and an n-type semiconductor, a three terminal configuration is adopted (see Fig. 24) so that the tunneling tip injects electrons into the metal film, which is held at ground potential. The current, I_c, that passes through to the substrate is measured. As an injector, the tip bias is negative, and we refer to its magnitude as V_t. For spectroscopy, the tip bias is swept while the tip–surface distance is varied to maintain a constant injection current.

Figure 55 shows I_c versus V_t curves for Au–Si(111) and Au–GaAs(110) heterojunctions grown in UHV and measured in dry N_2 at atmospheric pressure (Bell and Kaiser, 1988; Kaiser and Bell, 1988). The sharp onset at ~ 0.8 eV for Au/Si(111) gives the value of the SBH while multiple onsets can be extracted from the Au/GaAs(110) data. The simplest BEEM models use a one-dimensional (1D) theory

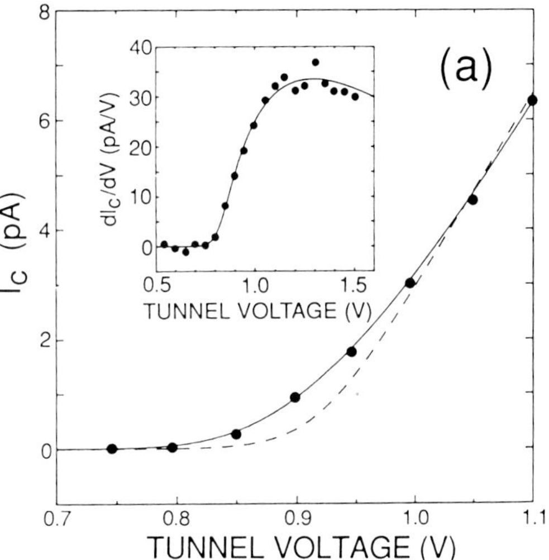

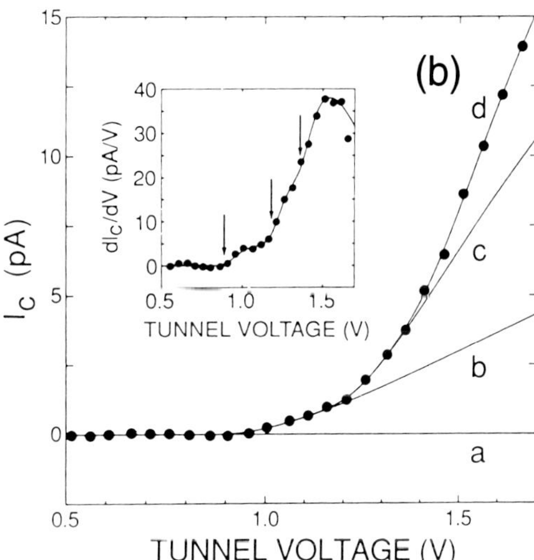

Figure 55. I_c versus V_t spectra for (a) Au–Si(111) and (b) Au–GaAs(110) near the threshold region for heterojunctions grown in UHV and measured in dry N_2 at atmospheric pressure. In the Au–Si(111) spectrum, the dotted line is a fit to a one-dimensional transport theory. The four Au–GaAs(110) spectra were taken at different positions on the same sample; note the apparent linearity in the curves beyond threshold, although the magnitude of I_c varies drastically. The insets present derivative spectra from which the structure related to the semiconductor band structure is evident (Bell and Kaiser, 1988).

to describe electron transmission across the interface. The differences between 1D theory and experiment are noticeable in Fig. 55 around the threshold regions. These differences can be modeled using the actual semiconductor conduction band shapes while maintaining a free-electron picture for the metal film. Therefore, since energy and transverse momentum must be conserved at the interface (assuming no scattering), the momentum–matching conditions governed by the band structure of the semiconductor become crucial. The Au–Si(111) interface has a single conduction band minimum at Γ to produce a single onset; the shape beyond the threshold is modified by inclusion of the band structure. In contrast, Au–GaAs(110) has three conduction band minima which clearly appear in the BEEM dI_c/dV_t spectra in the inset. The lowest onset at 0.89 eV corresponds to the accepted SBH value and is due to the CBM of GaAs at Γ. The onsets at 1.18 eV and 1.36 eV indicate additional channel for electron transmission across the interface due to conduction band minima at the L and X points, respectively.

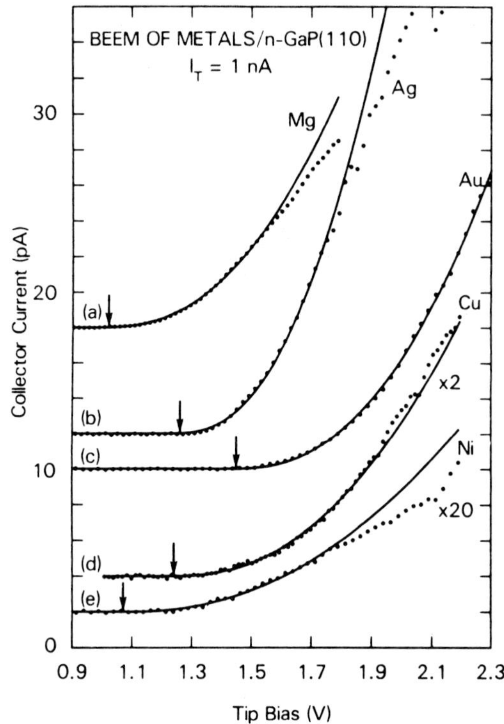

Figure 56. I_c versus V_t spectra for Mg, Ag, Au, Cu, Ni and Bi on GaP(110) indicating the dependence of the SBH on a specific metal (Prietsch and Ludeke, 1991).

Dependence of the SBH on specific metals has been demonstrated with BEEM in a set of experiments performed completely in vacuo on GaP(110) (Ludeke et al., 1991; Prietsch and Ludeke, 1991; Prietsch et al., 1991). Figure 56 indicates the existence of thresholds specific to Mg, Ag, Au, Cu, Ni and Bi in good agreement with other data, particularly when the fits were improved by including scattering and image effects in greater detail. While roughness or a lack of coherence across an interface affects the transmission function, inelastic scattering in the metal film removes electrons from the phase space that may cross the barrier giving an $I_c \propto (V - V_0)^{5/2}$ dependence at higher values of V_t. Electron–electron scattering dominates over phonon and impurity scattering and these effects were modeled here by including an exponential decaying functional form the mean free path. These improvements in the electron transport model were shown to give fits with lower SBH threshold values than those obtained from more elementary theory (Bell and Kaiser, 1988).

New studies of BEEM at high biases suggest interesting carrier transport properties across a metal–semiconductor interface (Ludeke, 1993). For the Cr/GaP(110) system, it has been reported that high bias operation can produce bulk density of states features of GaP in BEEM spectra. Besides two thresholds which represent different conduction band minima, the magnitude of I_c on the smooth part of the sample for tunneling biases >7.5 eV was found to exceed the injection current and structures at ~ 4 eV and ~ 6 eV, as shown in Fig. 57. In addition, spectral features were observed independent of position or the magnitude of I_c. The electron multiplication effect was attributed to impact ionization in the semiconductor. The energy dependence of inelastic scattering in either the metal or the semiconductor is generally structureless. Large amounts of scattering would reduce the forward focusing and preclude large multiplications of I_c reducing the effects of electron multiplication in the metal. In fact, the mean free path of the electrons is such that scattering would be expected to take place near the interface. Eliminating other possibilities, such as interferometric effects in the metal layer, scattering in the semiconductor was considered the most likely origin of current intensification. Taking advantage of the fact that the Cr/GaAs(110) interface is disordered, relaxing the momentum conservation condition, the observed spectra at

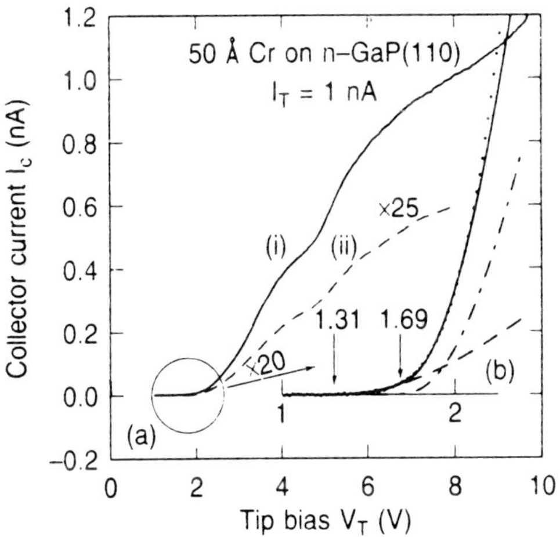

Figure 57. (a) I_c versus V_t for the Cr/GaP(110) interface (i) on a smooth section, (ii) on a rougher section; (b) is the expanded and magnified edge region of (a) (Ludeke, 1993).

high biases were assigned to GaP conduction band structure. To support this assignment, calculations of the conduction band of GaP were shown to agree with the data, suggesting that final state effects in the semiconductor can be probed with BEEM at high biases.

The Au/GaAs(110) spectra in Fig. 55 indicate that different slopes are observed beyond the SBH threshold (Kaiser and Bell, 1988). This is due to position-dependent electron transmission coefficients across the interface, while the I_c versus V_t threshold value typically remains fairly constant. When the tip is scanned above the sample at fixed V_t and I_t, forward-focusing allows high spatial-resolution imaging of variations in I_c versus (x, y) to give a picture of inhomogeneities in a buried interface. Such imaging is typically done in conjunction with STM imaging of the metal film surface to determine whether surface morphology is related to the interface morphology. Any mechanism that enhances disorder reduces I_c to a far greater extent than it affects the spatial resolution on account of the forward-focusing condition. It has been inferred from this observation and from detailed comparisons of STM surface images and BEEM images of the buried interface that topography at the interface provides more image contrast than differences in SBH (Ludeke et al., 1991). Topographic images of Au/GaAs(100) and Au/AuAl/GaAs(100) films are fairly smooth, the BEEM image of the Au/GaAs(100) interface is considerably less homogeneous than that of the Au/AuAl/GaAs(100) interface (Kaiser et al., 1989). For these systems, it is known that Au diffuses readily at the interface unless an AlAs diffusion barrier is present. For the latter case, therefore, there is a greater area of the interface that supports electron transport. We also note that oper-

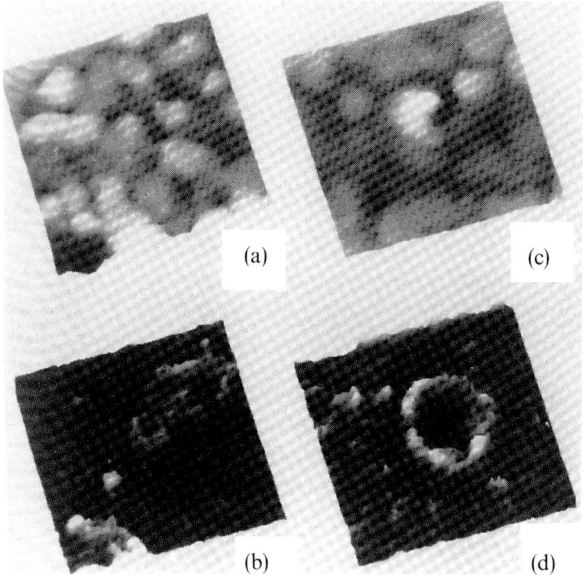

Figure 58. 800 × 800 Å^2 STM (a, c) and BEEM (b, d) images of Au/Si(111). For the STM images, the gray scale covers a vertical range of 40 Å and 58 Å, respectively. In the BEEM images, the gray scale covers a current range of 2–37 pA and 0–33 pA, respectively. Images (a) and (b) correspond to the as-prepared interface. Images (c) and (d) are the result of interfacial modification after application of high biases (Fernandez et al., 1990).

ation in the high-bias regime has indicated that irreversible electron-induced changes can occur; this was demonstrated for the Au–Si(111) system prepared by various methods (Fernandez et al., 1990).

Following successful implementation of BEEM, a complementary spectroscopy, involving ballistic holes created by direct injection or by a scattering process in the base region, was demonstrated (Bell et al., 1990; Hecht et al., 1990). This work required increased sensitivity and low temperatures compared with the BEEM measurements previously described. In the previous discussion we have considered BEEM (negative tip bias) with *n*-type collectors where a (positive) electron current is measured provided the ballistic electrons surmount the potential barrier at the base–collector interface. While it was noted that a reverse bias might be expected to result in a null collector current, we recall that the BEEM experiment involves a base and collector at zero bias. Since the reverse bias condition refers to electron current from sample to tip, holes created in the bare region may scatter with equilibrium electrons. Then, hot electrons that are produced can flow to the collector and be detected as a (positive) current. This situation is depicted with its complement, a *p*-type collector, in Fig. 59. For a *p*-type collector, a positive tip bias results in the injection of ballistic holes. A negative collector current can result in electron–electron scattering in the base and the creation of hot holes which can be collected. Holes follow analogous spectroscopic and focusing conditions to electrons. It should be noted that ballistic electrons can enter the collector only well above threshold, so that the hole current cannot be assigned to any process involving the semiconductor conduction band.

2.6 Metal Surfaces

Metal surfaces are considerably "smoother" electronically than semiconductor surfaces because of the delocalized nature of metallic bonding. Even if local-

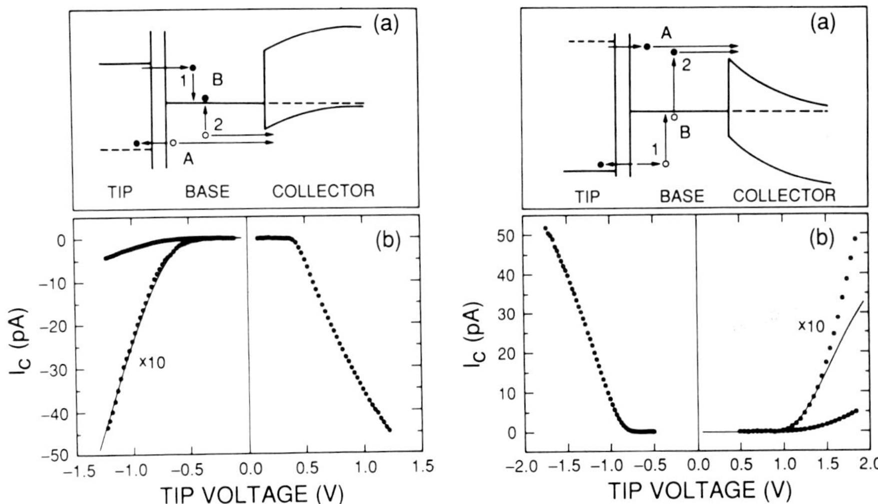

Figure 59. Left: diagrams and spectra for BEEM (a) and for electron–electron scattering processes (b) for a *p*-type semiconductor collector. For a positive tip voltage, BEEM injects ballistic holes, while, for negative tip biases, electrons may scatter in the base, creating a distribution of hot holes which may be collected. In the spectra of Au/Si(111), the negative current reflects the collection of holes. Right: diagrams and spectra for BEEM (a) and for electron–electron scattering processes (b) for an *n*-type semiconductor collector. As in the previous discussion for a negative tip bias, BEEM injects ballistic electrons into the base and these may be collected. A positive tip bias may create holes in the base region which can scatter with equilibrium electrons to create hot electrons that are collected. In the spectra of Au/Si(111), the positive current reflects the collection of electrons (Bell et al., 1990).

ized, d-orbitals extend some short distance into the vacuum, the delocalized electrons in s-p orbitals at the surface tend to smooth the charge contour at energies near E_F that dominate the states that form the tunneling image. Thus, corrugation amplitudes on metal surfaces, as imaged by even a sharp STM tip, may be expected to be rather small, reducing the prospect of obtaining atomic resolution. Improvements in the mechanical and electronic stability of STMs have allowed considerably progress in imaging metal atoms at surfaces, however. Even without atomic resolution, STM can be used to probe the structure and motion of steps (Wolf et al., 1991), or to analyze surface diffusion; these issues are closely related to growth processes. The earliest success at atomic-resolution imaging of a metal surface was reported for the Au(110) surface which exhibits "open" terrace structures (Binnig et al., 1983a). At that time, a missing row model with (111) facets along the $[1\bar{1}0]$ direction was proposed based on STM images of local 1×2 and 1×3 structures. Atomic-resolution images of close-packed surfaces were later reported and these were found to exhibit unexpectedly large corrugations in some cases. The study of adsorbate-induced interactions on metal surfaces, where surface reconstructions are removed or promoted by the presence of the adsorbed species, and the study of vicinal surfaces to elucidate step dynamics directly affect growth issues. As an extension to spectroscopic measurements previously discussed, potentiometric measurements have begun to relate the role of grain boundaries to macroscopic resistivity in metals.

2.6.1 Close-Packed Surfaces

The first atomically-resolved images of a close-packed metal surface were obtained on the Au(111) surface both in air and in ultrahigh vacuum (Hallmark et al., 1987a). The image in Fig. 60a taken with a tunneling current of 3 nA at a tip bias of +30 mV illustrates the close-packed surface structure of a flat terrace of Au(111) imaged in UHV. A corrugation amplitude of ~0.3 Å denoted Au atoms separated by ~2.9 Å, and monolayer steps separating (111) terraces. Atomic-resolution images of the surface were also obtained in air and beneath organic monolayers deposited onto the surface. For images obtained in vacuum, processing of the tip by over-biasing was found to improve the lateral resolution considerably, but tunneling instabilities occurred fairly often. Imaging in air or through an ex-situ deposited organic monolayer provided a dampening medium between tip and sample which reduced these instabilities. The unexpectedly large corrugation (between 0.05 Å and 0.3 Å) on this close-packed surface was partially attributed to operation with a relatively low tunneling resistance, i.e. scanning very close to the surface. Furthermore, an electronic enhancement was proposed due to the existence of a localized, occupied surface state near E_F.

The surface of Al(111) was also found to produce clear atomically-resolved images (Wintterlin et al., 1989) with relatively large corrugations. A representative image with a corrugation of the order of 0.3 Å is shown in Fig. 60b; similar images were obtained for a range of biases between −1 V and +1 V. Again, the corrugation heights were improved with a treatment of over-biasing with related instabilities in the tunneling current during scans. Analyzing images obtained during the tip preparation procedure indicated that the tip was lengthened by ~25 Å as a result of over-biasing, while the surface beneath the tip became quite rough. This tip-induced improvement originates with atoms trans-

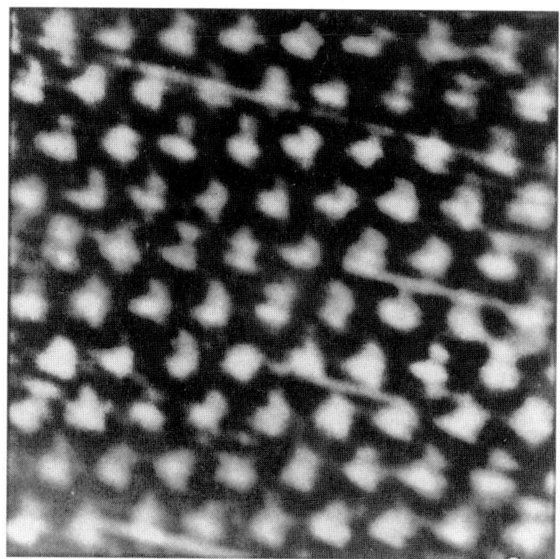

(a)

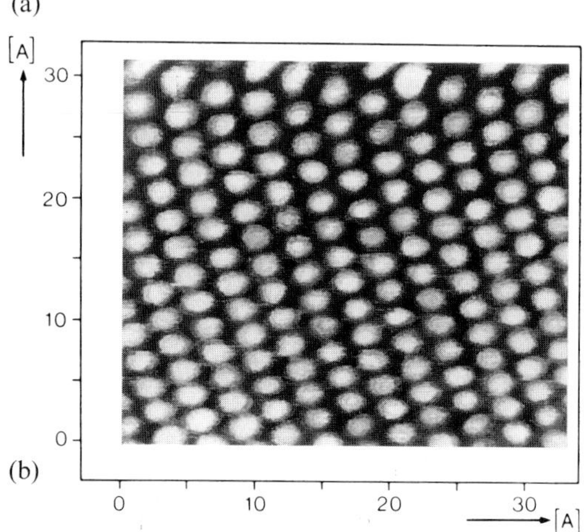

(b)

Figure 60. (a) 12 × 14 Å² image of clean Au(111) thin film measured in ultrahigh vacuum with a tip bias of +30 mV and a tunneling current of 3 nA. The atoms are approximately 3.0 Å apart, and the measured corrugation under these conditions was 0.3 Å (Hallmark et al., 1987a). (b) 34 × 34 Å² image of an Al(111) surface measured in ultrahigh vacuum with a tip bias of −50 mV and a tunneling current of 6 nA. The measured corrugation under these conditions was 0.3 Å (Wintterlin et al., 1989).

ferred to restructure the tip on account of strong tip–surface interactions. Besides these changes, it was postulated that the surface is further deformed by a tip–surface attractive force with a resulting enhancement in corrugation.

Although more pronounced in the case of semiconductors, electronic effects have been shown to enhance images on metal surfaces. In fact, both localized surface electronic states and tip–surface interactions with elastic deformation have been considered as the origin of enhanced corrugation for atomic resolution imaging of close-packed metal surfaces. In support of the enhanced corrugation observed on Au(111) surfaces, a state of ~0.3 eV below E_F was observed in spectroscopic measurements of Au (Everson et al., 1991). This state was strongest on terraces and reduced in intensity, though not significantly shifted in energy, at steps and at sputtering-induced pits. Theoretical investigations (Ciraci et al., 1990) used first-principles calculations for an Al tip atom held some distance above an Al(111) surface. Considering three distance regimes – far, near, and contacting – it was found that, at sufficiently close distances, lateral confinement of current-carrying states enhances lateral resolution. In particular, p_z-like states are induced on the metal by the presence of the tip. In fact, in imminent contact, the corrugation may actually be inverted due to tip-induced states (Tekman, 1990). These results suggested that elastic deformation may not lead to enhanced corrugation because, when the site-dependent deformation is compensated by the feedback system of the STM, corrugation in the image is actually de-enhanced (Ciarci et al., 1990). Another theoretical view of image enhancement beneath contamination layers involved the role of resonance-tunneling (Zheng and Tsong, 1990).

Combined STM and low energy ion scattering studies were performed on the single-crystal intermetallic NiAl(111) alloy surface (Niehus et al., 1990) in an effort to resolve the long-standing question as to whether the surface is terminated by Ni or Al atoms. Prior measurements had shown that a mixture of Ni and Al domains exist for a (1×1) surface structure, as determined by LEED. In these studies, step heights, terrace, and island boundaries were used to determine whether mixed terminations exist. Samples annealed at ~1100 K exhibited flat terraces separated by double steps with several small, azimuthally-oriented islands of ~20 Å diameter. These islands, representing a minority surface species, exhibited only a single step height. On the basis of grazing-incidence ion scattering, it was determined that the predominant surface species was Ni. With this information, the STM images led to the important conclusion that the minority islands consist of Al with the Ni beneath. Higher annealing temperatures that resulted in the total removal of residual oxygen contamination were found to remove the Al residing at the surface entirely. For more complicated systems, as treated in this study, the necessity of combining analytic techniques to obtain both structural and compositional information is clear.

2.6.2 Surface Diffusion

Diffusion of metal atoms at surfaces represents a first step in considering the dynamic processes involving mass transport or the motion of steps. Observations of the self-diffusion of metal atoms on metal surfaces in real time was first demonstrated for the Au on Au(111) system (Jaklevic and Elie, 1988). Here, individual atoms were not observed, but time-lapse imaging provided a view of the evolution of tip-induced defects on the surface at room temperature, proving that a significant amount of diffusion occurs at this temperature. Figure 61 shows a series of time-lapse topographic images. Images (1) shows a 400 Å $\times$ 400 Å scan of the surface where vertical monatomic steps are observed. Image (2) was taken after the tip was touched to the surface twice, forming two craters 2–3 atoms deep and involving a loss of ~ 3000 atoms. Higher resolution images showed that these craters were aligned with low-index step directions with the initial healing process taking place fairly rapidly. The rest of the image sequence taken over ~2 h follows the evolution of the craters by diffusion at room temperature. As depressions fill in and mounds smooth out, the shapes of the craters do not necessarily conform to the lowest-index steps. This indicates that the rate-determining process is the capture of atoms from remote regions of the surface. Quantitative estimates of the room temperature Au surface diffusion coefficient of 8.4×10^{-15} cm^2/s were obtained (Drechsler et al., 1989) with STM by a quantitative analysis of the evolution of Au microhills on an Au surface.

2.6.3 Stepped Surfaces

Before achieving atomic resolution on metal surfaces (Hallmark et al., 1987a), the step structures of the Au(111) surface were investigated (Kaiser and Jaklevic, 1987) on a microscopic scale. This microscopic complement to techniques (e.g. diffraction) that average over a large area was shown to be crucial for the determination of the energetics of metal surfaces on an atomic scale. By measuring the distribution of terrace sizes and step heights, it was established that the equilibrium Au(111) structure involves ordered arrangements of steps which might be explained by repulsive step–step interactions (Kaiser and Jaklevic, 1987).

This approach was carried over to a reconstructed Pt(100) surface, where the relation between domain boundaries and surface steps was explored (Behm et al., 1986). The Pt(100) surface undergoes an irreversible reconstruction at $T > 420$ K to

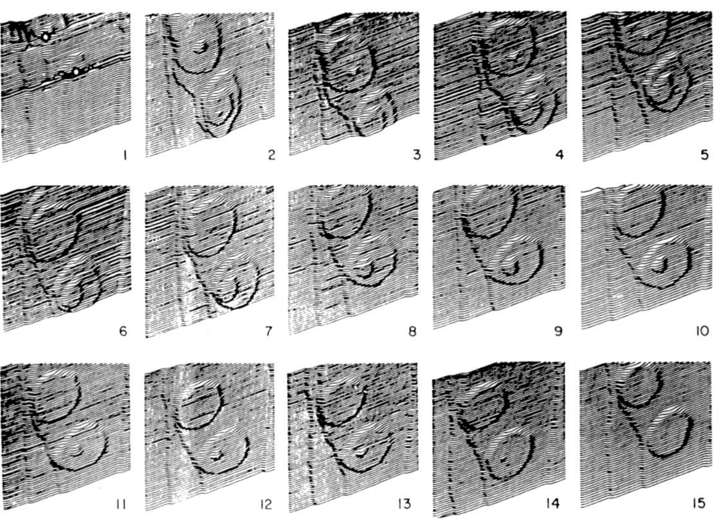

Figure 61. Series of time-lapse (time/frame = 480 s) 400×400 Å^2 STM images of the Au(111) surface at room temperature. Steps are one atomic unit in height. In Frame 1, the two circles indicate the positions where the tip was touched to the surface. Frame 2 shows the 2- and 3-atom deep craters left after contact. The remaining frames show that the smaller craters are filled by diffusing atoms over an ~2 h period (Jaklevic and Elie, 1988).

a 5×25 "hexagonal structure", as determined by LEED. STM images revealed a reconstructed Pt layer above an unreconstructed (1×1) second layer producing a surface corrugation with a height of ~0.4 Å and a repeat distance of ~14 Å (five-fold periodicity). Images across steps, as shown in Fig. 62, indicated that the reconstruction extends into the step so that the step is the grain boundary between the reconstructed lower terrace and the unreconstructed second layer of the upper (reconstructed) terrace. It was found that only one reconstructed domain can exist on an individual terrace (which might extend for >1000 Å). This implies that the existence of different rotational domains on a single terrace is energetically unfavorable. The relative energetics of the step–domain interaction and the domain–domain interaction were further inferred from observations that steps do not preferentially orient the reconstructed domains, nor do adjacent domains appear to interact.

It is reasonable to assume that repulsive interactions between steps exist in systems where steps are found to be spaced fairly far apart. In view of these repulsive interactions, finite temperature leads to thermal disorder and the production of kinks, which is involved in the so-called step roughening transition. Recent studies (Frohn et al., 1991) of the step distance distribution on vicinal Cu(100) surfaces of the (1 1 n) type showed a peak at a value large enough to avoid the formation of (1 1 1) microfacets, but smaller than that expected for a purely repulsive interaction, as shown in Fig. 63. Here it is seen that the repulsive interactions stabilize the (1 1 7) surface, but the (1 1 19) surface exhibits peaks for two terrace width distances. An additional longer-ranged attractive interaction over ~3–5 atomic distances, as from parallel oriented dipole moments, was proposed as

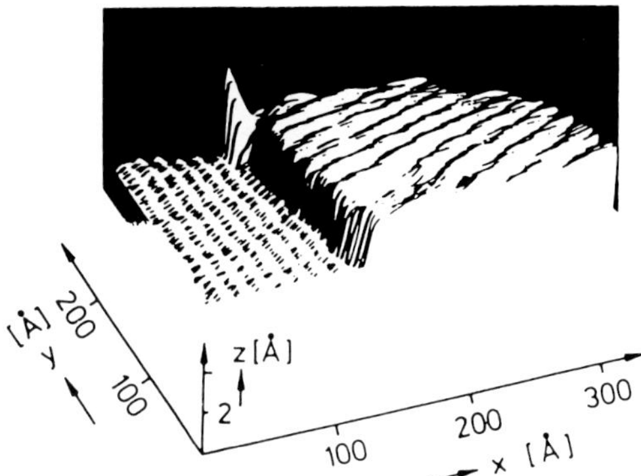

Figure 62. The hexagonally-reconstructed Pt(100) surface showing that two uncorrelated domains are separated at a surface step. The reconstruction extends to the step (Behm et al., 1986).

a mechanism that can result in the observed step distance distribution.

With variable temperature capabilities, a direct measurement of step roughening has also been made with STM (Wolf et al., 1991). Imaging steps on the vicinal Ag(111) surface at room temperature produced

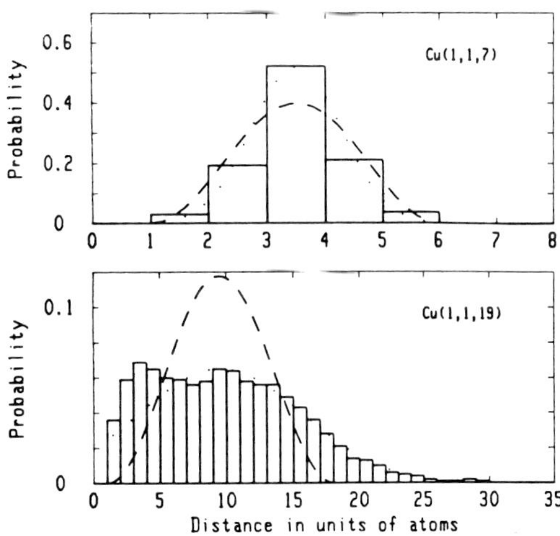

Figure 63. Distance distributions for Cu(1 1 7) (top) and Cu(1 1 19) (bottom). The (1 1 7) is stabilized by the repulsive step–step interaction, as modeled in the dashed curve. However, attractive interactions produce faceting on the (1 1 19) surface and the repulsive model (dashed curve) does not apply (Frohn et al., 1991).

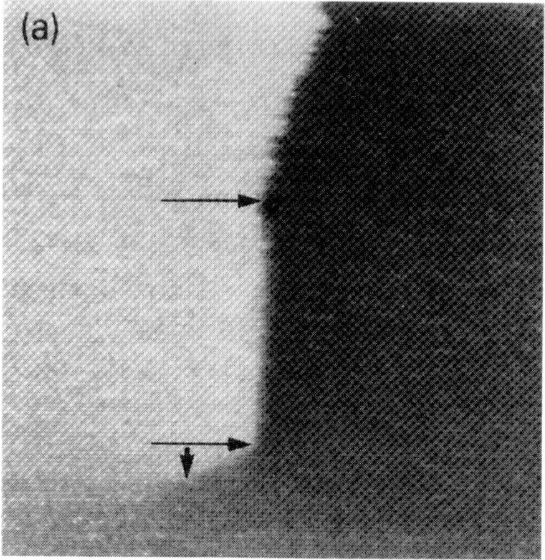

(b)

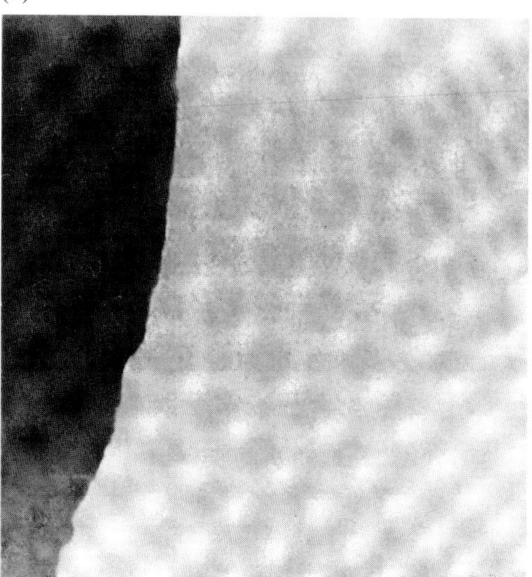

Figure 64. (a) A 340×340 Å^2 image of Ag(111) in the vicinity of a step that ends in a screw dislocation (short arrow) measured at room temperature. Between the two long arrows, the step runs along a ⟨110⟩ direction and is one layer in height. Frazzling is observed above an impurity (upper arrow) due to the diffusion of kinks, but the impurity blocks this diffusion to the lower kink-free region. (b) A 340×340 Å^2 image taken at 27 K showing a sharp step with no evidence of kink diffusion (Wolf et al., 1991).

images with a frazzling of steps along the scan direction. This is seen in Fig. 64a, where the ends of two steps occur at a bulk screw dislocation, as previously imaged for Cu(111) (Samsavar et al., 1990). The frazzling suggests that, at room temperature, kinks can diffuse along monatomic steps on the time scale of a scan. At room temperature impurities were found to reduce or eliminate the frazzling, effectively reducing the rate of emission and recapture of adatoms at steps. For contamination-free surfaces at 27 K, the frazzling was gone, demonstrating that the diffusion of kinks and even the motion of signal adatoms is essentially frozen, and this step sharpness could be maintained up to ~ 150 K.

2.6.4 Adsorbate-Induced Reconstructions of Metal Surfaces

It is well-known that, besides forming ordered overlayers, adsorbates can modify surface structure either by removing clean-surface reconstructions or by inducing the formation of ordered reconstructions involving the topmost metal layer(s). Since the strength of adsorbate–metal bonds can be comparable to metal–metal bonds, it is reasonable that the surface can be disrupted drastically by adsorption. Mass-transport, or more simply the density of the topmost atomic layer, is a crucial issue. If a process requires significant mass transport, the question arises as to how an entire terrace can be transformed at relatively low temperatures (e.g., ~ 300 K), or whether a model requiring only local atomic rearrangements should be chosen. Since averaging techniques monitor periodicity over a relatively large area, information regarding the uniformity of a reconstruction at the level of individual unit cells can only be obtained by a local probe like STM. Likewise, assessing residual disorder on the atomic scale requires atomic scale imaging. We consider three systems to illustrate adsorbate-induced removal of a reconstruction [CO/Pt(110)] (Gritsch et al., 1989), promotion of a reconstruction through the adsorbate interaction with mobile atoms supplied from steps [O/Cu(110)] (Coulman et al., 1990), and promotion of a reconstruction involving the local removal and rearrangement of surface atoms [O/Cu(100)] (Jensen et al., 1990). We note that specific adsorbate superstructures can stabilize the structure of steps on vicinal surfaces, as for the S/Cu(11 1 1) system (Rousset et al., 1989).

LEED studies have established that the clean Pt(110) surface exhibits a (1×2) reconstruction that transforms to a (1×1) structure upon adsorption of molecular CO; the transformation can occur for temperatures as low as 250 K. If this structural modification represents a transformation from a missing-row structure to a surface layer in registry with the bulk, 0.5 ML of Pt atoms must be transported from sources such as step edges. Since a large amount of mass transport is an unexpected possibility for a low temperature structural transformation, two alternative models were put forth on the basis of LEED studies. In the first, a sawtooth reconstruction was proposed where only small atomic displacements would be required. A second model was based on the occurrence of an order–disorder transition. The STM images in Fig. 65 show that (A) CO transforms the unexposed (1×2) surface layer into (B) a

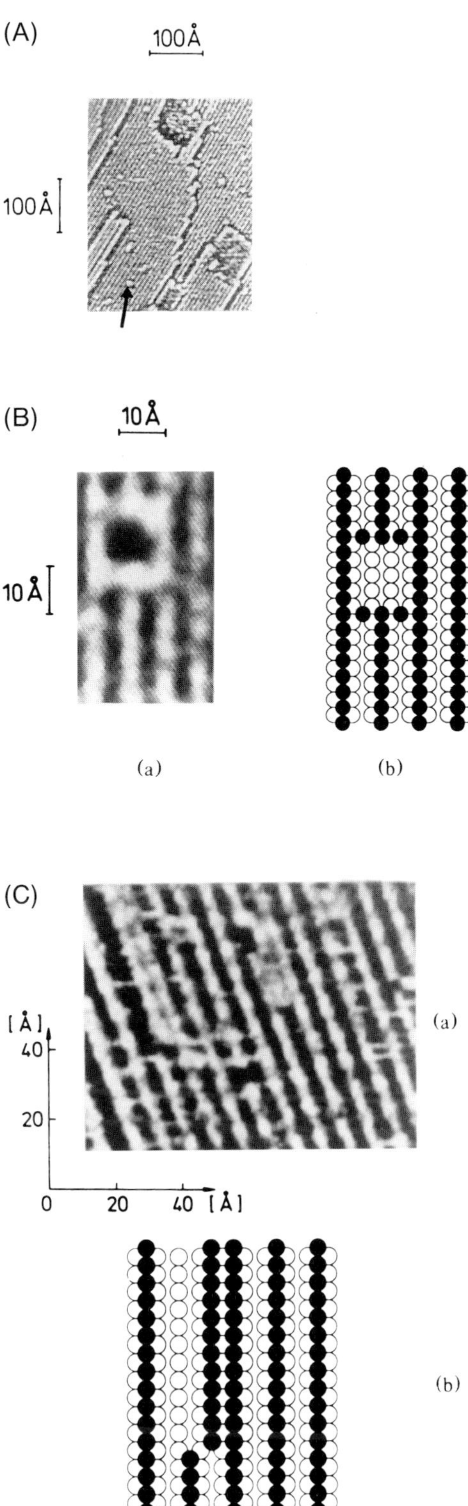

Figure 65. (A) An STM image of the clean Pt(110)-1×2 surface in ultrahigh vacuum. (B) The appearance of holes (a) after 1.5 L exposure to CO and a detailed model (b) of the formation of a hole. (C) An STM image (a) taken at 350 K showing a more extensive growth of a 1×1 structure (b) due to CO exposure (Gritsch et al., 1989).

series of holes and rings for a room temperature exposure and (C) begins the transformation to a 1×1 surface for exposure at 350 K. This microscopic picture provided by STM indicates that a homogeneous nucleation process occurs with local transformations to form a 1×1 topmost layer involving atomic motion over only a few lattice sites. At room temperature a surface atom overcomes an activation barrier to begin the formation of a locally-ordered structure; at higher temperatures there is a correlated movement of atomic rows to form long-range quasi-1×1 order.

The local rearrangement of the CO/Pt(110) system can be contrasted to the long-range ordering that occurs for the O/Cu(110) system. The atomic structure of the (2×1)O–Cu(110), as observed by STM, consists of a series of long Cu rows along the [001] direction, as shown in Fig. 66, which implies a correlated rearrangement of the surface. One controversial aspect regarding the formation of this structure at ~ 300 K is whether significant mass transport is involved (Chua et al., 1989), or if the structure is formed from "added-rows" of Cu atoms atop the unreconstructed substrate (Coulman et al., 1990). Careful observations of the step structure showed that the steps retract as the (2×1) islands condense, indicating that the oxygen induces the formation of new Cu rows – an "added-row" reconstruction – with a different local surface density to the topmost layer of the clean surface. The proposed mechanism to form linear steps involves trapping diffusing Cu atoms through coadsorption with O, which stabilizes the structure. The strong Cu–O–Cu interaction, with O adsorbed at bridge sites, stabilizes linear chain formation, while weak attractive forces between rows cause the nucleation of adjacent chains. This can be viewed as the precipitation of

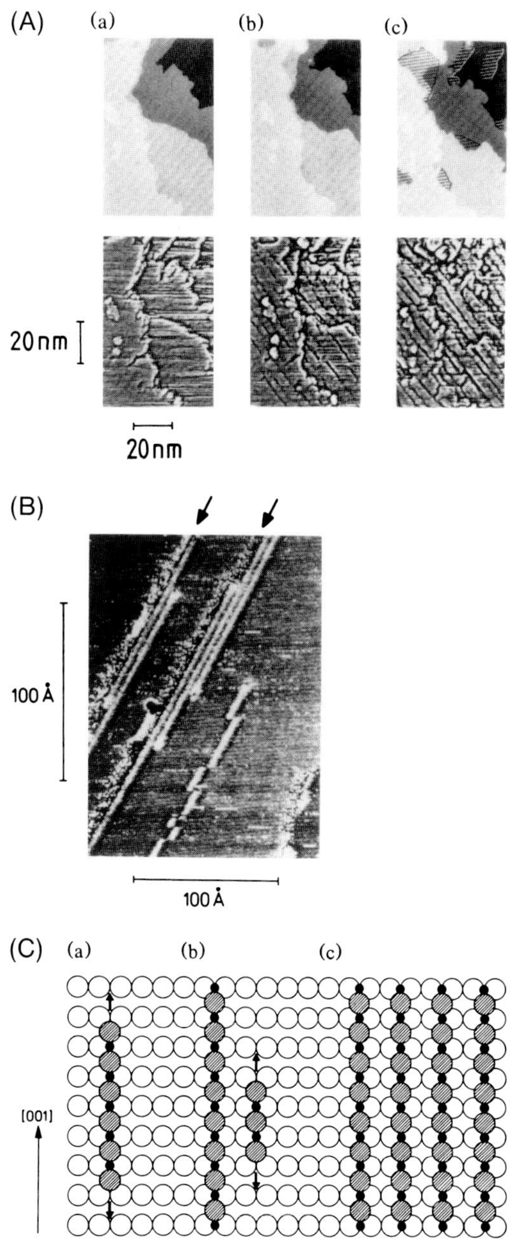

Figure 66. (A) STM images of (a) the clean Cu(110) surface, (b) the Cu(110) surface after 6 L oxygen exposure, and (c) the Cu(110) surface after 10 L oxygen exposure. The upper panels show the step-terrace topography and the hatching indicates where the topmost layer has been removed. (B) The "added-row" structure and (C) an atomistic model showing the preferential growth of long oxygen-stabilized (black circles) rows (Coulman et al., 1990).

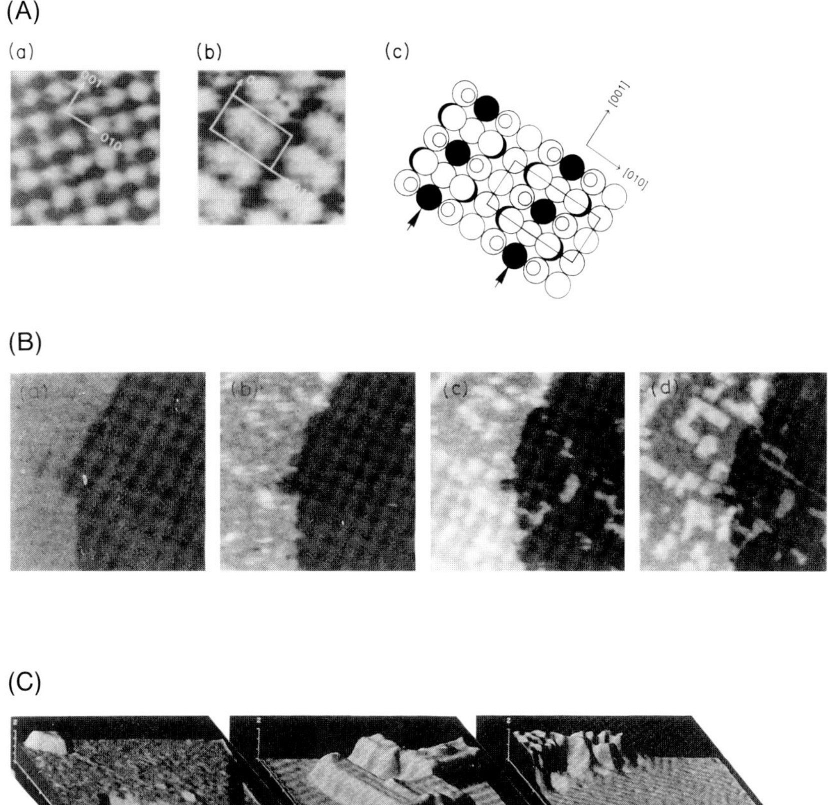

Figure 67. (A) Atomic-resolution images of the oxygen-induced $(2\sqrt{2}\times\sqrt{2})\,R45^\circ$ structure on C(100). (B) The formation of the reconstructed phase after room temperature exposures to oxygen from clean (left) to saturation exposure (right). (C) Images after annealing the room-temperature oxygen-saturated surface at 300 °C (left and center). The regions between the islands and the tops of the islands are reconstructed. Disorder occurs after annealing at 500 °C (right) (Jensen et al., 1990).

a solid from a fluid mixed phase in two dimensions.

A third type of nucleation at elevated temperature has been observed for the O/C(100) system (Jensen et al., 1990). As seen in Fig. 67 the (1×1) Cu surface is transformed to a $(2\sqrt{2}\times\sqrt{2})\;45^\circ$-O structure at room temperature. The mechanism for this reconstruction involves the local motion of Cu atoms, in which one of four Cu[001] rows are removed and the displaced atoms nucleate and grow epitaxially atop the Cu surface in a disordered phase. The reconstructed structure consists of Cu–O–Cu chains along the [001] direction, as for Cu(110), but the reconstruction grows at the expense of the removal of 25% of the surface atoms. Upon annealing, the disordered Cu phase becomes ordered with the reconstruction also occurring atop the Cu islands.

2.6.5 Growth of Metallic Adlayers

As for the case of epitaxial growth on semiconductor surfaces, the details of homo- or hetero-epitaxial metal-on-metal growth depends on the structure of the underlying substrate and the energetics of the atom–atom interactions between and within the atomic layers. Following previous discussions on STM observations on metal systems, i.e., adatom diffusion, step dynamics, and adsorbate-induced reconstructions, the Ag-on-Au(111) system may be taken as an example of epitaxial metal-on-metal growth from the sub-monolayer to multilayer regime (Dovek et al., 1989). Clean Au(111) exhibits a ($p \times \sqrt{3}$, $p = 22$–23) reconstruction, as illustrated for an atomically-flat terrace in Fig. 68 a. The gentle corrugation of the reconstruction repeats approximately every 70 Å. 0.25 ML Ag deposition results in the formation of epitaxial islands that conform to the clean Au reconstruction (Fig. 68 b). Development of a full Au monolayer is seen (Fig. 68 c) to remove the reconstruction entirely, and this is followed by the growth of Ag in clusters (Fig. 68 d) that conform to the symmetry of the Ag(111) surface being created. Upon mild annealing, the epitaxial layer exhibits a surface roughness no greater than that of the initially clean substrate.

2.6.6 Resistivity in Polycrystalline Metals – Scanning Tunneling Potentiometry

Electrical resistivity in crystalline materials originates from collisions between the current carriers and non-periodicity in the lattice potential. When polycrystalline materials are considered, the possibility of scattering at potential steps, such as at grain boundaries, must also be taken into account. Thus, the microscopic basis for resistivity in metals relies on the distribution of scattering sites where carrier chemical potential gradients exist. Quasi-simul-

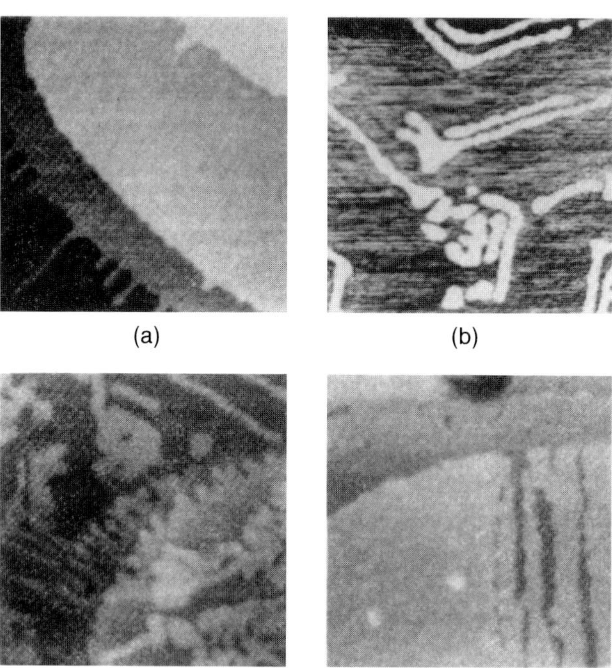

Figure 68. 1150×1150 Å^2 STM images of Au(111) after evaporation of (a) 0.1 ML Ag, (b) 0.25 ML Ag, (c) 0.75 ML Ag, and (d) 1.5 ML Ag. At lowest coverages, Ag conforms to the ($p \times \sqrt{3}$) reconstruction of the underlying layer, but this reconstruction is no longer observed above one monolayer (Dovek et al., 1989).

taneous structural and potentiometric imaging, i.e., using the STM as a potentiometer, were performed to determine the relationship between potential steps in a metallic film and its structural morphology. This methodology, initially developed (Muralt et al., 1987) to measure the potential distribution across a semiconductor heterojunction, was refined to study polycrystalline $Au_{60}Pd_{40}$ (Kirtley et al., 1988 b). The scanning tunneling potentiometry experiment is illustrated schematically in Fig. 69 A. Operation is in an interrupted-feedback mode with the sample bias pulsed with a square wave between the imaging bias and zero. For the potentiometry measurement, a potential difference is placed across the sample, causing a current to flow, and the potentiometer is set at a given location, such that the tip–sample bias is zero. As the tip is scanned away from the null position, the sample potential, V_S, is obtained using the relation $V_S = I_0 R$, where I_0 is the tunneling current and R is the tunneling resistance. Operating near a zero-tunneling-bias condition improves the signal-to-noise ratio over the previous method (Muralt et al., 1987) and allows a relatively large potential difference to be placed across the sample. The spatial correlation of the terraces with the surface potential of the grains in the topographic image is clearly seen in Fig. 69 B. The major regions where potential drops occur is at grain boundaries (over ≤ 1 Å) while potential gradients within a grain occur probably due to phonon scattering, as inferred from temperature-dependent measurements (Kirtley et al., 1988 b).

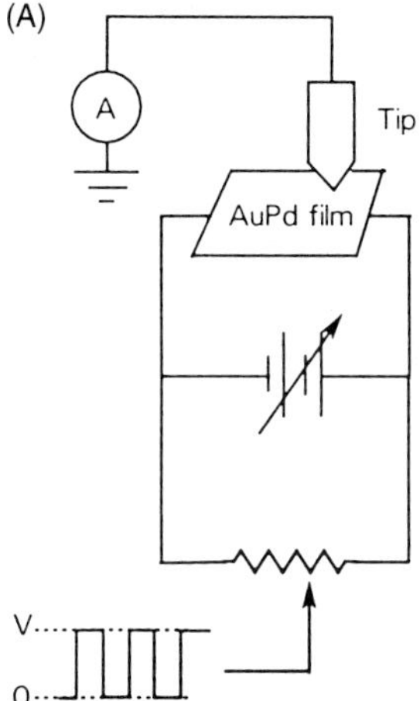

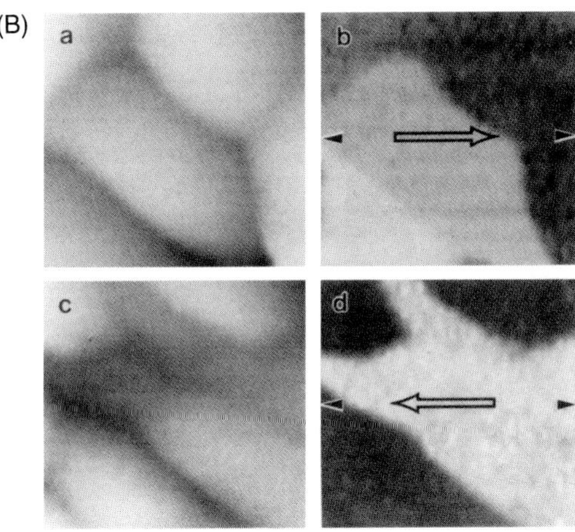

Figure 69. (A) Diagram of scanning tunneling potentiometry experiment. (B) Topographic (left) and potentiometric (right) images of an $Au_{60}Pd_{40}$ film under fields of (b) +85 V/cm and (d) −85 V/cm (Kirtley et al., 1988 b).

2.7 Insulators

The operating principle of STM requires electrical conduction through the material being imaged. At first glance, imaging insulators and wide-band-gap semiconductors without surface pre-treatment, such as by depositing a conducting template layer, would be considered to be difficult if not impossible. However, as discussed in this section, STM imaging and spectroscopy can be performed on these surfaces with near-atomic resolution. Such analysis provides an important complement to other techniques for the formulation of the structural and electronic bases of chemistry at insulating surfaces related to, for example, catalysis, electrode–electrolyte interactions and metal–ceramic bonding. While atom-resolved structural information could alternatively be obtained using AFM (Sec. 3), local electronic structure can only be determined by STM. Imaging single crystal and polycrystalline materials with wide bulk band gaps typically requires higher tunneling biases so as to reach electronics states for tunneling. As previously noted in the case of the CaF_2 system (Avouris and Wolkow, 1989 b), the position of the Fermi level with respect to the conduction or valence band edge is crucial to determine biasing conditions. In addition, surface states may exist in the bulk band gap providing states for tunneling and site-specific imaging. Furthermore, doping these materials often provides sufficient electron transport for imaging. While these states allow lower bias voltages, the general complication of ensuring surface cleanliness and ideal stoichiometry exists for many of these systems. The use of complementary techniques to analyze composition (in vacuum) is one means of alleviating some of these complications. It should be pointed out that the major problem in imaging polycrystalline materials is in the ability to distinguish topographic and spectroscopic variations, particularly at grain boundaries.

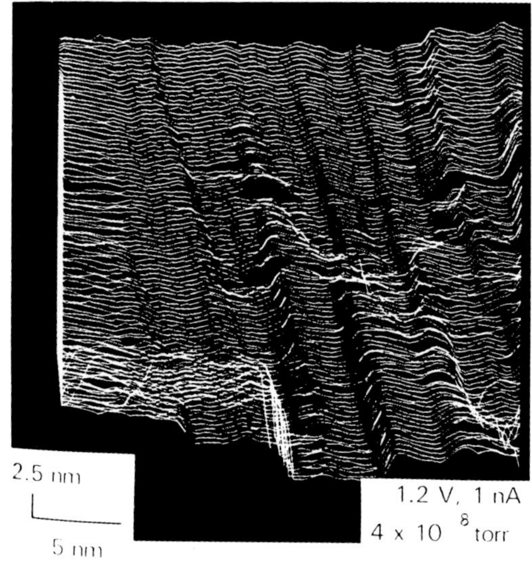

Figure 70. A 250 × 250 Å^2 image of an SiC basal plane obtained in a vacuum of 10^{-8} torr (Bonnell, 1988).

The first success in imaging the surface of a bulk insulator was the study of a single crystal of SiC by STM in air and in vacuum (Bonnell, 1988; Bonnell and Clarke, 1988). Figure 70 a shows a 250 × 250 Å^2 image (in vacuum) of the SiC basal plane subsequent to etching in HF. Steps separated by ~14.5 Å and varying in height between 14 Å and 30 Å were observed. $I-V$ curves obtained in air showed an asymmetric, or rectifying, behavior, while those obtained in vacuum were more symmetric. This was attributed to contamination on the air-exposed surfaces. Three states were found to exist in the SiC bulk band gap (2.89 eV) for spectra acquired in vacuum. Whether these states are intrinsic, originate from contamination due to etching, or signify a surface oxide was not determined.

Transition metal oxides, such as the ionic wide-band gap semiconductor TiO_2, have a wide variety of applications including use as catalyst supports and as electrodes. In these applications physics and chemistry on the atomic-scale are very important. Likewise, in the formation of interfaces, the chemistry of bonding at the surface is important. Several STM images of TiO_2 obtained in air (Fan and Bard, 1990; Gilbert and Kennedy, 1990) or in the presence of an electrolyte (Itaya and Tomita, 1989) have been reported. STM images of reduced TiO_2(110) (Rohrer et al., 1990) obtained in vacuum illustrate most clearly the complex surface topographies exhibited by these systems. Figure 71 shows an unoccupied state image ($V_{tip} = -2$ eV) of TiO_2 obtained in UHV after in vacuo cleaning, transfer through air, and in situ annealing. Periodic rows oriented along [1$\bar{1}$1] and spaced by 8.5 Å are observed; the corrugation along the rows exhibits a periodicity of 3.4 Å. Dark rows about 20 Å wide and along [$\bar{1}$13] run parallel to surface steps; their heights varying between 2 Å and 10 Å. These images could not be directly related to a bulk-terminated structure. Analysis of the images required consideration of surface non-stoichiometry related to an oxygen deficiency plus crystallographic shear. In the proposed model, electrons tunnel into unoccupied Ti states such that the light regions are attributed to Ti atoms and the dark regions correspond to O atoms. Inclusion of a $\frac{1}{2}$[0$\bar{1}$1] shear

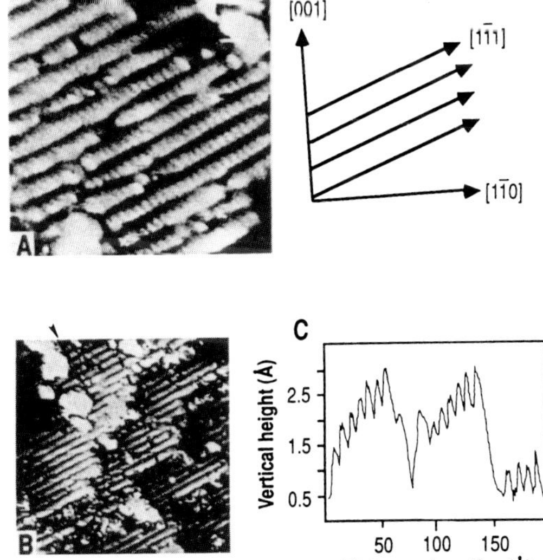

Figure 71. (A) A 109 × 109 Å^2 image of a TiO_2(110) surface showing rows along the [1$\bar{1}$1] direction. (B) A 246 × 246 Å^2 image over a region containing steps long [$\bar{1}$13] with heights of ~0.5–5 Å. The steps are illustrated by a profile (C) along the dotted line in (B) (Rohrer et al., 1990).

component provides an explanation for the observed surface structure, where the observed row corrugations are steps 1.6 Å high.

Imaging polycrystalline materials presents special problems with respect to single crystal surfaces because the presence of grain boundaries makes the definition of crystallite axes difficult. Furthermore, the variation in conductivity between individual grains and across grain boundaries must be distinguished from the structure. Recent studies of polycrystalline ZnO have addressed the topographic and electrical information that can be obtained at grain boundaries crystallite axes (Bonnell, 1988; Rohrer and Bonnell, 1990), as well as the effect of chemistry in imaging these surfaces (Bonnell and Clarke, 1988; Rohrer and Bonnell, 1990; Rohrer and Bonnell, 1991). Bismuth oxide is often added to ZnO to improve the characteristics of this material as a varistor. Increased resistance at grain boundaries is attributed to segregated Bi which bends the adjacent ZnO bands upwards and creates a depletion region. The images of polycrystalline ZnO, as shown in Fig. 72 exhibit large flat areas attributed to individual grains of ZnO; these regions, which comprise ~90% of the image, contain steps and facets. Between these flat regions are depressed areas depicting the grain boundaries; the areas appear depressed due to their low electrical conductivity. Some metallic regions were attributed to impurities. On the grains, tunneling spectra indicate that the surface Fermi level is pinned between two defect-induced surface states. The bulk Fermi level is typically pinned near the top gap by shallow donor states. It is found that the measured gap in the low conductivity region is considerably narrower near the grain boundary rather than farther away. These variations in the band gap, which

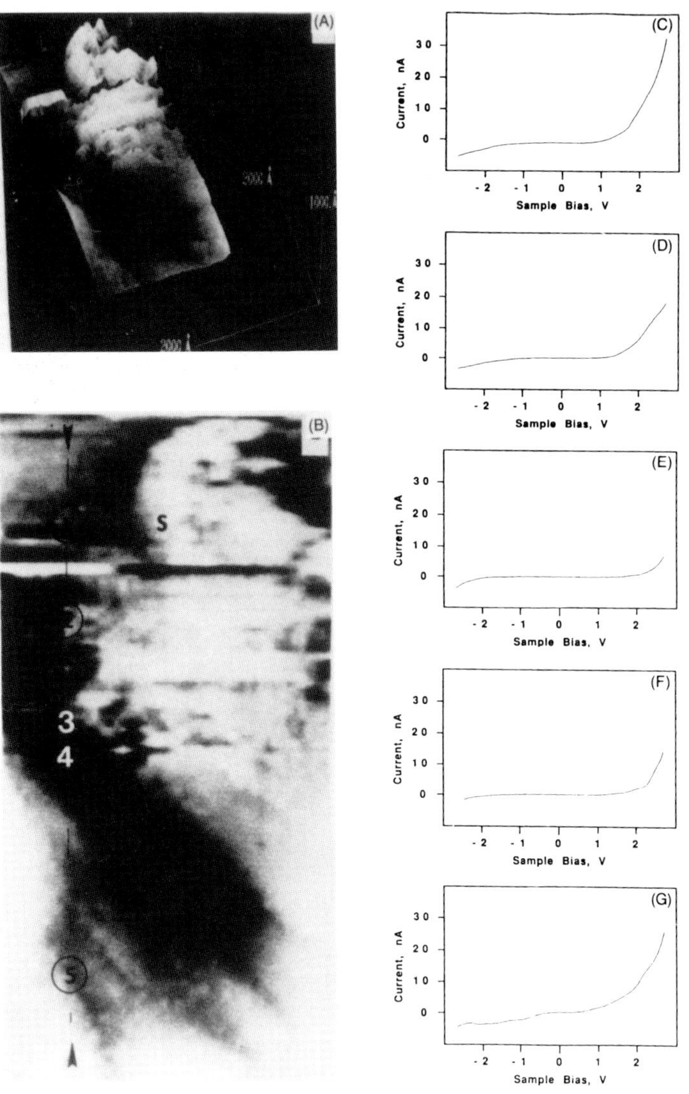

Figure 72. (A) A wider range image and (B) a 2000×4000 Å^2 image of polycrystalline ZnO centered at a grain boundary. The bright regions correspond to ZnO crystallites. $(dI/dV)/(I/V)$ curves for five locations along the grain boundary are shown; position 1 ⇒ C, position 2 ⇒ D, position 3 ⇒ E, position 4 ⇒ F, position 5 ⇒ G (Rohrer and Bonnell, 1990).

was found to be composition-dependent, were attributed to the sort of local band bending that increases the resistivity of the polycrystalline material. This conclusion assumed that the variations in the band gap originated from the bulk and were not solely a surface phenomena.

$SrTiO_2$ and other perovskite metal oxides are important in surface photocatalysis and as a substrate for thin films. The 2×2 ordered superstructure of the $SrTiO_2(100)$ surface observed by LEED and measured by photoemission is thought to comprise an ordered array of surface defects consisting of oxygen vacancy complexes which give rise to a mid-gap electronic state. STM studies (Matsumoto et al., 1992) show that both the undoped and the Nb-doped $SrTiO_3(100)$ surface exhibit flat regions with a random array of adsorbed particles after annealing at 1200 °C. The undoped sample exhibits conductivity due to the vacancy-induced state in the band gap. Figure 73 shows the surface of a Nb-doped sample in a region where a step separates three flat regions. The 4.2 Å step height comprises a unit cell height, so the termination of the two terraces is the same (either SrO or TiO_2). Regions A and C are slid by one-half period, while regions A and B are rotated with respect to each other. These atom-resolved images actually reveal an 8×8 Å^2 square lattice rotated 26° with respect to the bulk crystal axis; the corrugation amplitudes are similar to those observed for a metal surface. The lattice size corresponds to a 2×2 superstructure reported by LEED. Calculations suggest that the geometric and electronic structures can be explained with the surface comprised of a TiO_2 top layer with a 2×2 arrangement of (oxygen-vacancy$-Ti^{3+}-$O) complexes. The observed adsorbed particles might correspond to Sr atoms or SrO complexes.

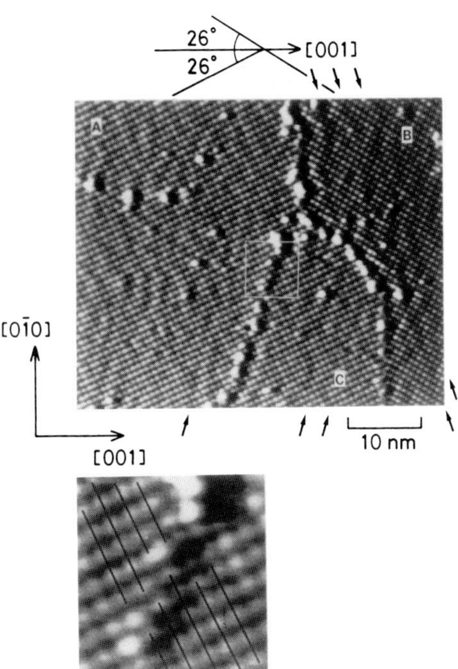

Figure 73. A 400×400 Å^2 current image of the $SrTiO_3(100)$ surface obtained with a tip bias of +0.6 V and tunneling current of 400 pA. The step separating the three regions of square lattices is clearly observed as well as the half lattice constant slip between regions A and C and a 52° rotation between regions A and B (Matsumoto et al., 1992).

2.8 Layered Compounds

Graphite is one of the most commonly used materials in conjunction with STM. Because of the ease of preparation of a flat, clean surface, graphite is used as a substrate for deposition of metallic, organic, or biological thin films and for calibration of the lateral scale of the piezoelectric elements. Since graphite is a layered compound exhibiting weak interlayer bonding strength, preparation is as simple as exposing a fresh surface layer by peeling upper layers off with adhesive tape. Since the surface is fairly inert, it was demonstrated very early on that the graphite surface could be imaged in air (Schneir et al., 1986). Many ideas related to STM and STS, as well as the interaction of tip and surface, were developed from early images of graphite, as discussed below. There has also been considerable interest in the relationship between the geometric and the electronic structure of intercalated graphite, and it is natural to compare such systems with pure graphite. Furthermore, layered compounds in general exhibit excellent tribological properties since individual layers can slide over each other fairly easily. A future goal is to relate the geometric structure between layers to the macroscopic frictional forces.

The early experimental and theoretical work on graphite indicated that the interpretation of images was less straightforward than anticipated, although atomic resolution is rather easy to achieve (Schneir et al., 1986). The STM image of

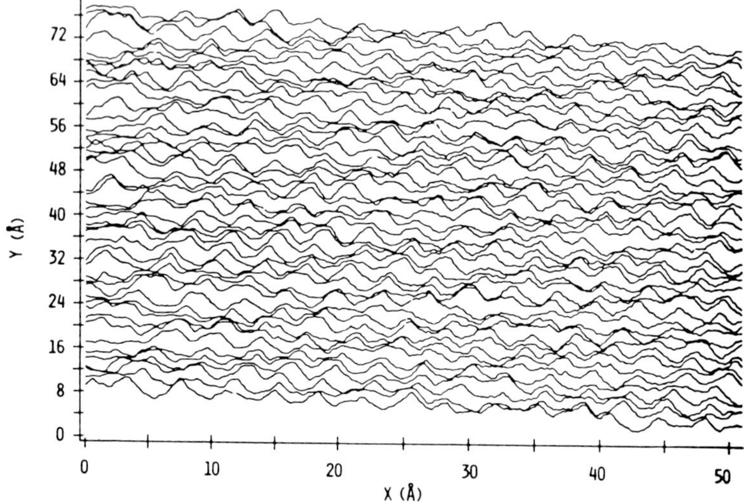

Figure 74. Linescans of the (0001) surface of pyrolytic graphite recorded with $V_t = 55$ mV and $I_t = 1$ nA. The regular maximum corrugation is ~3.5 Å (Batra et al., 1987).

graphite (0001) in Fig. 74 taken with $V_t = 55$ V and $I_t = 1$ nA in air already hints at some puzzling aspects. First, the images exhibit three-fold structures, although the lattice is hexagonal. Second, the measured corrugation is ~3.5 Å; this is significantly greater than the calculated corrugation of the π^* states (0.2 Å) or electronic states near E_F (0.8 Å) (Selloni et al., 1985). In addition, a variety of asymmetries can appear in the atomic images. The left panels of Fig. 75 show that an entire array of anomalous images may be observed depending on the tip conditions.

Observation of threefold rather than sixfold symmetry is explained by interlayer interactions that create a narrow but finite Fermi surface (Selloni et al., 1985). The ABAB stacking sequence of graphite, as shown in Fig. 76, identifies two types of C atoms: α and β. α atoms sit atop atoms in the second layer while β atoms do not. Theoretical calculations (Batra et al., 1987; Tománek et al., 1987) of the electronic structure of the graphite surface that the β states dominate the STM images, corresponding to those near E_F.

Explanations of the enhanced corrugations have taken into account the unique

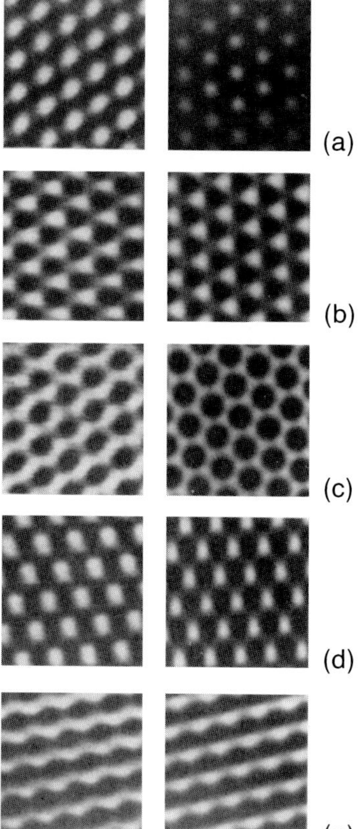

Figure 75. The left panels show experimentally-obtained images of graphite, while those on the right correspond to computer-generated images, as discussed in the text. The experimental image in (a) is the nominal graphite image (Mizes et al., 1987).

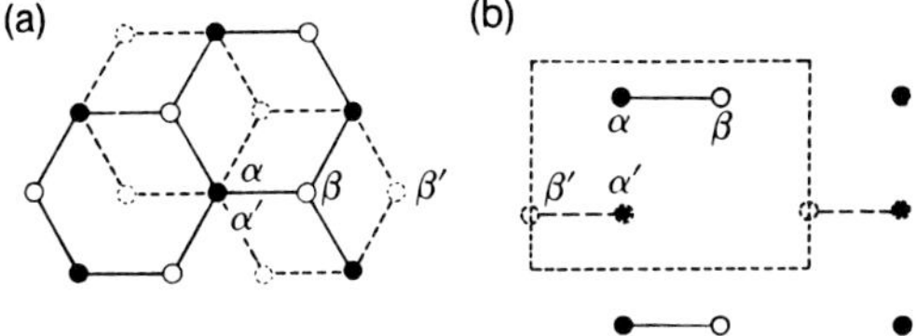

Figure 76. The two successive layers of hexagonal HOPG graphite showing the inequivalent α and β sites; (a) top view and (b) side view (Tománek et al., 1987).

electronic effects of low-dimensional systems (Tersoff, 1986), the role of tip–surface interactions (Soler et al., 1986), and the effects of contamination in the tunnel gap (Mamin et al., 1986). In principle, all may apply to some extent. For systems of low dimensionality, the possibility exists for the Fermi surface to collapse to a point at the corner of the surface Brillouin zone (SBZ) (Tersoff, 1986). For tunneling biases where only a single state is involved in the tunneling, a single wavefunction forms the image. If $\boldsymbol{k}$ is at the SBZ edge and the crystal potential has a Fourier component $2k_F$, a gap will form at E_F with the formation of standing waves with nodes. This conditions can be met by graphite, a zero-gap semiconductor, as well as by charge density wave systems (Coleman et al., 1985) and semiconductor surface states (Stroscio et al., 1986) either because $2k_F$ is the reciprocal lattice vector of the surface or because of a $2k_F$ charge density wave distortion. Then the image is a trace of the individual electron wavefunction; while the image may have the periodicity of the lattice, it does not necessarily reflect the true atomic positions. Also, mixing the states with the layers below, while weak, produces some modification to the simplified description of a single row or sheet of atoms. This would reduce the corrugation amplitude with respect to the idealized row or sheet.

Besides electronic effects, the corrugation on graphite increases with decreasing V_t and increasing I_t. While a low V_t is a condition for imaging near E_F, which can produce strong corrugations due to electronic effects, these conditions suggest that the effect is greatest when the tunnel gap is the smallest. The suggestion was therefore made that the interaction between tip and sample can produce elastic deformations in the materials in which the interplane bonding strength is weak, leading to the observed effect (Soler et al., 1986). Figure 77 presents the results of a parameterized calculation which shows the conditions under which the compression and expansion of graphite might be observed with respect to a rigid surface. We note that operating in the constant height node would not be expected to result in the observed

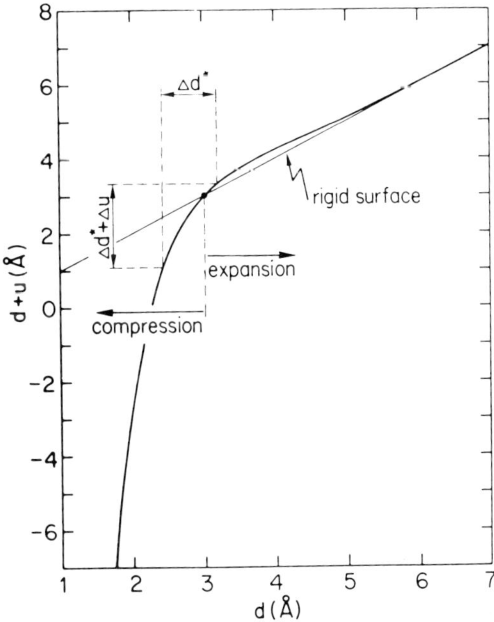

Figure 77. Results of a parameterized calculation for the deformation of graphite by the tunneling tip. The tip motion, Δd^*, produces an enhanced movement of the surface, $\Delta d^* + \Delta u$ (Soler et al., 1986).

tip-induced deformations (Tománek et al., 1987).

The variety of images obtained from the same sample shown in the left panels of Fig. 75 indicate a range of anomalous structures of the same lattice that can be obtained with the STM. This can be attributed to tunneling from multiple sites on the tip (Mizes et al., 1987; Ishiki et al., 1990). In the simplest picture, the parameter that determines the image relates to the relative phases of the s-wavefunctions participating in the tunneling. A graphite image with threefold symmetry can be determined by three Fourier components of equal amplitude. If two tip sites are operable, the STM image becomes the superposition of two images, taking into account the relative amplitudes and phases of the tunneling from the two tip sites. A tip for which multiple tip sites contribute to the tunneling process can be described by a superposition of s-waves. The computer-generated images in Fig. 75 correspond to a triangular lattice in which the three Fourier components are shifted in phase and amplitude. Starting from a single tip (a), the anomalous images correspond to one Fourier component having a relative phase shift of (b) $\pi/2$ or (c) π. If the amplitude of two of the Fourier components dominates the third, the image of (d) results; if one Fourier component is dominant, the image of (e) is produced.

A point contact mode in which current flows through a contamination layer has been considered as an alternative explanation for some of the effects observed when imaging graphite in air (Colton et al., 1987). Of course, in terms of the interaction of the tip with the surface, a fairly large contact area might result (Mamin et al., 1986), resulting in the observed weak dependence in tunneling current as a function of z-displacement of the tip. This contamination-mediated deformation was taken to account for the fact that atomic images might be observed even with z-displacements over ≤ 100 Å. In support of these ideas, spectroscopic curves using graphite-coated or graphite (pencil lead) tips indicated a reduced barrier height and functional dependence different from that expected for tunneling through a vacuum barrier (Colton et al., 1987). It was also shown that a tip composed of a graphite flake can be taken as a trigonally-symmetric multi-atom tip to produce anomalous STM images.

There has been considerable interest in graphite intercalation compounds (GICs) over the years because of their unique electronic, optical, and transport properties. Donor GICs, such as C_nM, where M is an alkali-metal, donate charge to the graphite host. Acceptor GICs, such as C_nX, where X is a metal chloride or other Lewis acid, remove charge from the graphite host. The repeat unit of intercalant layers within graphite layers is referred to as the stage; a stage-n GIC has an intercalant layer for every n graphite layers. Problems with contamination and stability of many of these materials have limited the study of their surface properties. Intercalants can be depleted in vacuum and are often very reactive in air. However, preparation of samples for surface analysis and, in fact, for in situ characterization can be done in pure, inert gas atmospheres. The types of questions that arise regarding bulk structure of GICs can be extended to the surface. Incorporation of intercalant species affects the shape of the Fermi surface and the electronic structure. These factors plus differences in the screening are expected to alter the STM corrugation. In addition, information on the stoichiometry, structure, and nature of the surface species may be obtained using STM.

STM studies of stage-1 C_6Li in an Ar atmosphere (Anselmetti et al., 1989) were found to include three types of terrace structures with different in-plane lattice constants, as shown in Fig. 78. The lattice constant of image (a) is 0.35 nm, corresponding to an incommensurate surface lattice with nearly close-packed Li. Images (b) and (c) represent commensurate superlattices. In (b) the lattice constant is 0.42 Å, which is expected for a $\sqrt{3} \times \sqrt{3}$ structure, while the lattice constant of 0.49 Å in (c) corresponds to a 2×2 structure. The bulk structure is a $\sqrt{3} \times \sqrt{3}$. The STM corrugation was found to be comparable to that for HOPG for $V_t \leq 200$ mV, and significantly less than that of HOPG for $V_t \geq$ 200 mV. These results were in partial agreement with theoretical calculation (Selloni et al., 1988), but a greater than predicted corrugation for low biases led to the conclusion that the sampled layer is Li and not C. For this stage-1 compound, it is expected that cleaving will leave some of the intercalant layer on the surface, and it was proposed that the high mobility of Li on the surface could result in the occurrence of the incommensurate and 2×2 structures, the latter being related to bulk C_8Li, as well as the bulk-like $\sqrt{3} \times \sqrt{3}$ structure.

STM images of C_8M (M = K, Rb, Cs) show the prevalence of 2×2 structures, as in the bulk (Kelty and Lieber, 1989). In addition, C_8Rb and C_8Cs also exhibit novel non-hexagonal one-dimensional superstructures (Anselmetti et al., 1990). The 2×2 superlattice for C_8Rb is seen in Fig. 79a. Three rotational $\sqrt{3} \times 4$ domains of these one-dimensional structures are shown in Fig. 79b; a single domain $\sqrt{3} \times 4$ superstructure of C_8Cs is shown in Fig. 79c. In addition, $\sqrt{3} \times \sqrt{13}$ superlattices were observed. The 2×2 appeared to have an enhanced corrugation possibly due to a charge density wave (Kelty and Lieber, 1989; Anselmetti et al., 1990). The one-dimensional structures have reduced corrugation and their presence is consistent with the depletion of alkali-metal atoms at the surface. Based on the domain sizes observed, the driving force for their formation is apparently short-ranged, and it was suggested that these effects may be caused by the collapse of the associated charge density wave (Anselmetti et al., 1990).

Studies of the acceptor stage-1 C_6CuCl_2 GIC produced polarity-dependent effects in the corrugation (Olk et al., 1990), as illustrated in Fig. 80. Images taken with $V_t = +10$ mV exhibited a uniform corrugation for β and α sites. However, images taken with $V_t = -10$ mV exhibited the α–β asymmetry of graphite. Adapting the interpretation of the calculations for C_6Li (Sel-

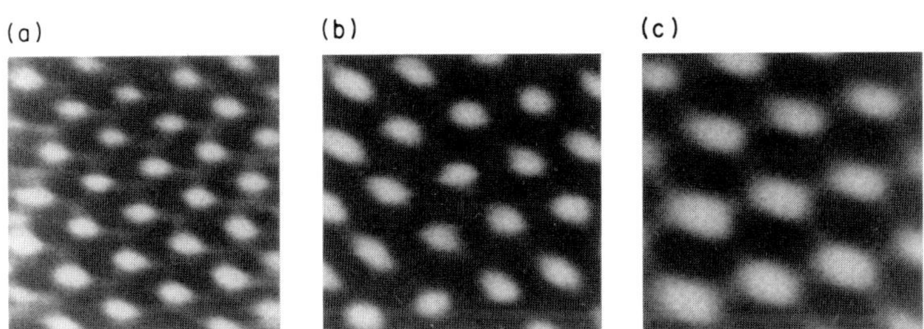

Figure 78. Three 18×18 Å^2 STM current images obtained from different terraces of a C_6Li GIC surface showing superlattices with lattice constants of (a) 0.35 nm (incommensurate), (b) 0.42 nm (commensurate) and (c) 0.49 nm (commensurate) (Anselmetti et al., 1989).

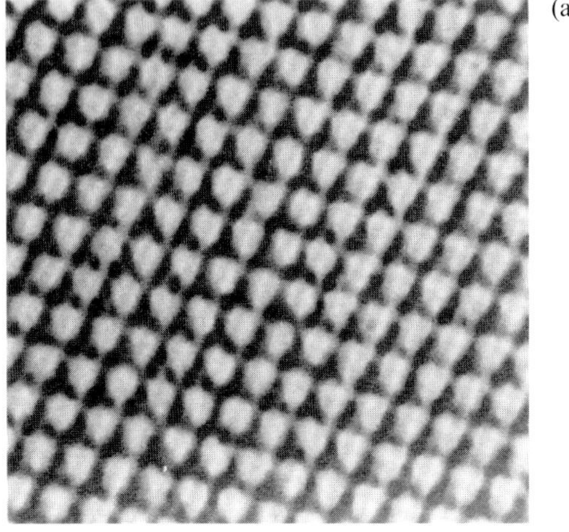
(a)

(b)

(c)

loni et al., 1988), it was concluded that $V_t > 0$ represents tunneling from $CuCl_2$ molecules beneath the first layer and $V_t < 0$ represents tunneling to the C atoms at the surface. The former conclusion indicates that the intercalant reduces or removes the interlayer coupling in graphite which results in the $\alpha-\beta$ site asymmetry. The removal of site asymmetry as a function of charge transfer and donor/acceptor type was also predicted by calculations of the electronic structure of GICs (Qin and Kirczenow, 1990).

A recent study of the tribological aspects of metal dichalcogenides demonstrates the complementarity of STM and AFM in addressing a problem of great technological interest by comparing and contrasting surface structural and electronic properties (Lieber and Kim, 1991). Layered materials are excellent lubricants because of the weak bonding interactions between individual planes. While the structures of many metal dichalcogenides are similar, MoS_2 possesses singularly superior lubricating capabilities. One of the goals of the study was to determine if a systematic understanding of substituted compounds could provide some insight into the uniqueness of MoS_2. STM and AFM images of the S atoms of MoS_2 show fairly uniform hexagonal structures with a period of 3.2 Å (Fig. 81 a). The Ni-substituted compound, $Ni_xMo_{1-x}S_2$ in Fig. 81 b indicates some changes in the corrugation of the S atoms that could signify differences in Ni–S versus Mo–S bonding, but no major differences in the surface structure between

Figure 79. (a) 50 × 60 Å^2 STM image of the 2 × 2 superlattice of C_8Rb ($V_t = -50$ mV and $I_t = 3.4$ nA); (b) 100 × 90 Å^2 image of three rotational $\sqrt{3} \times 4$ domains of C_8Rb ($V_t = -20$ mV and $I_t = 2.0$ nA); (c) 75 × 65 Å^2 image of a single $\sqrt{3} \times 4$ domain on C_8Cs ($V_t = 160$ mV and $I_t = 1.5$ A) (Anselmetti et al., 1990).

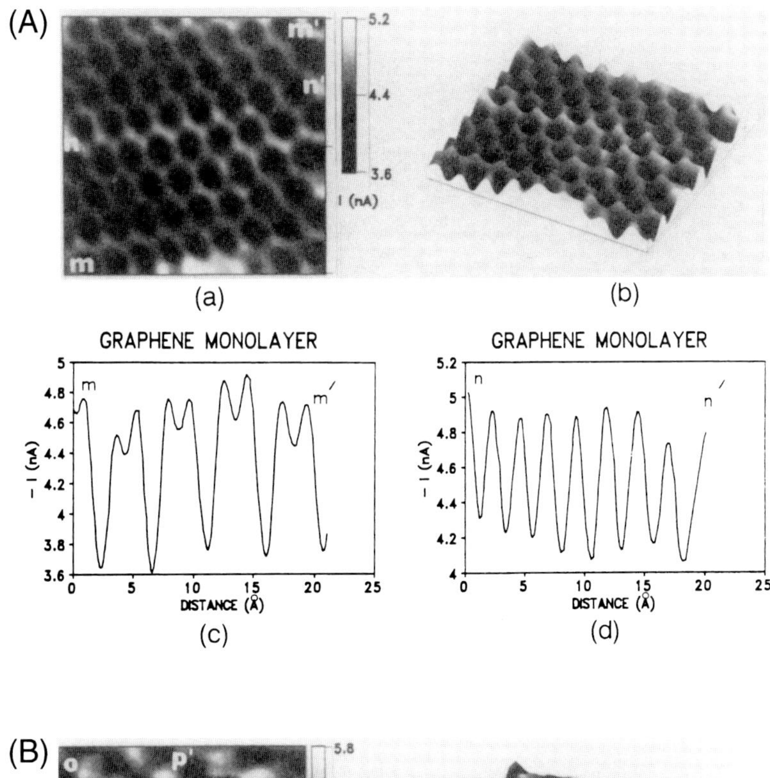

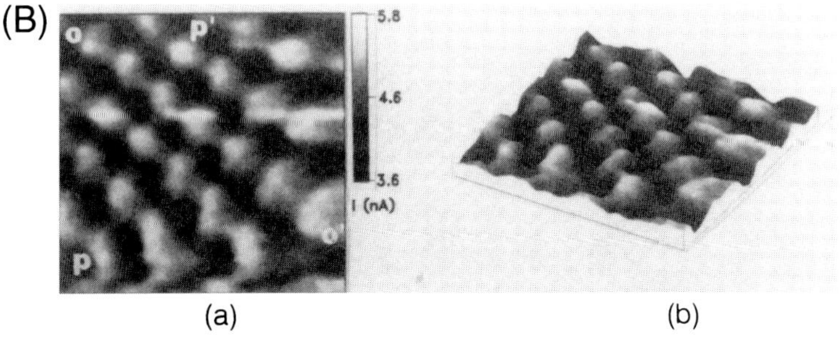

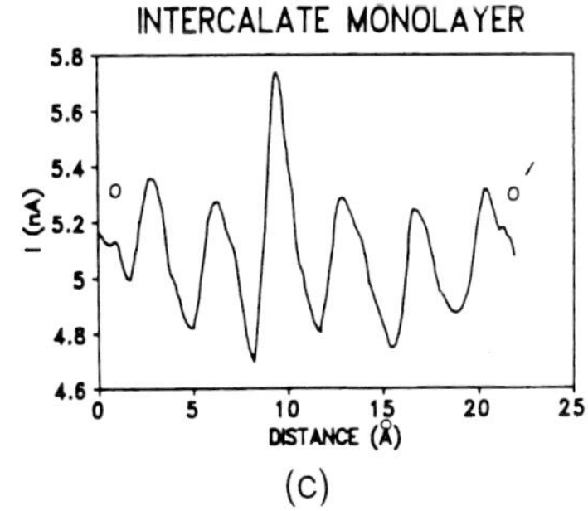

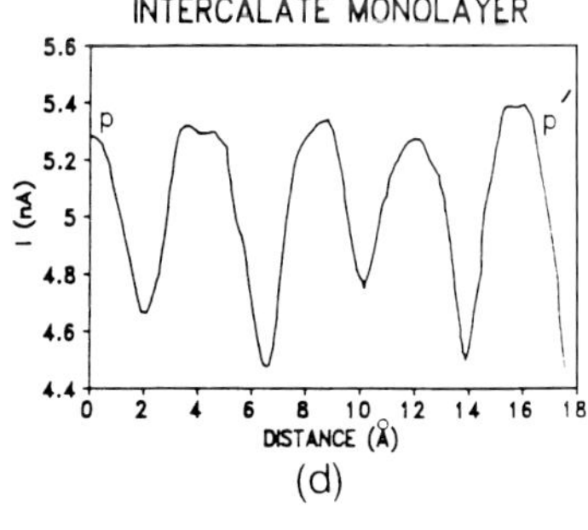

Figure 80. 20.75×20.75 Å^2 STM images of C_6CuCl_2 with line scans as indicated. (A) $V_t = 10$ mV; (B) $V_t = -10$ mV (Olk et al., 1990).

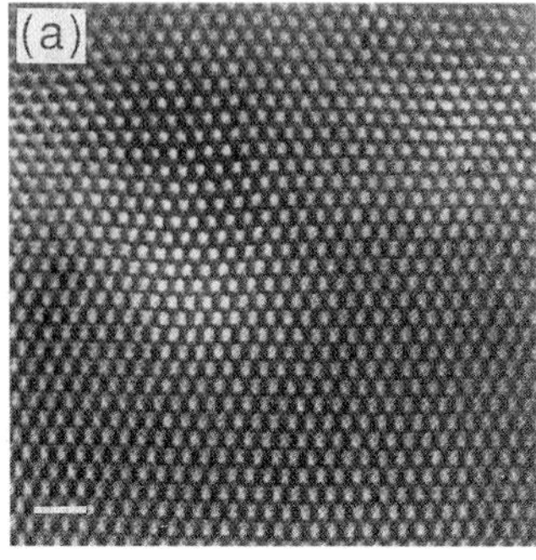

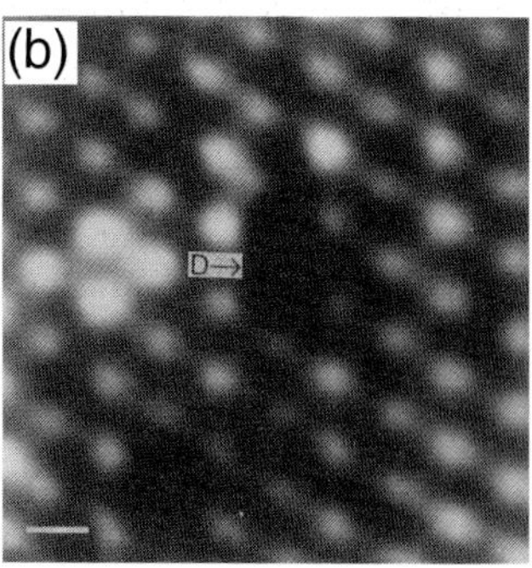

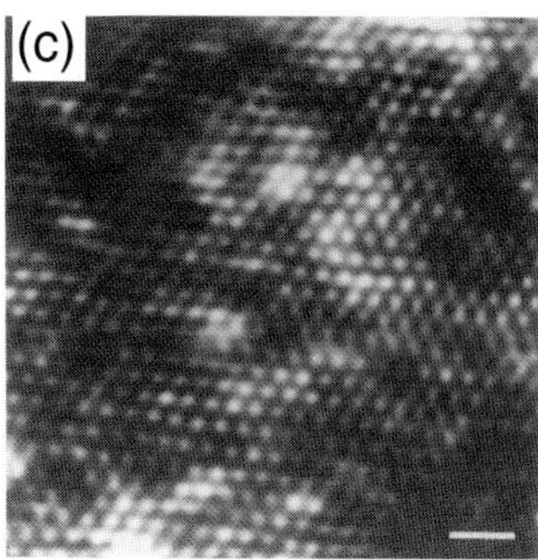

(d)

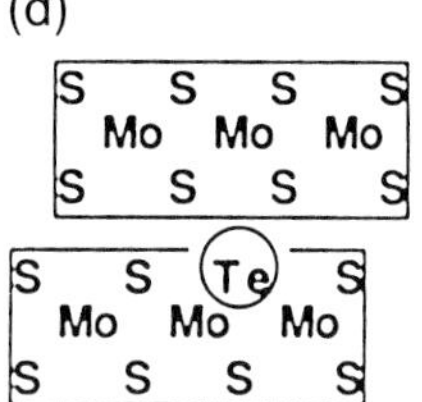

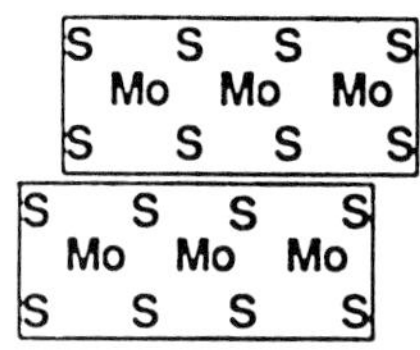

Figure 81. STM images of (a) MoS_2 (9×9 nm^2), (b) $Ni_{0.1}Mo_{0.9}S_2$ (2.5×2.5 nm^2), and (c) $MoS_{1.75}Te_{0.25}$ (8×8 nm^2). (d) Drawing depicting a possible structural modification in which the Te reduces the layer–layer interaction of pure MoS_2 (Lieber and Kim, 1991).

the two compounds. The STM images were significantly more striking. Localized electronic defects in states near E_F are signified by a marked reduction in the measured corrugation. Another compound, as seen in Fig. 81 c, Te-substituted MoS_2, $MoS_{1.75}Te_{0.25}$, exhibits atomic-sized protrusions, as observed by AFM. The number of such protrusions are scaled with the Te concentration. However, analogous localized electronic structure changes were not observed by STM. While no direct conclusions based on these observation have been made regarding the tribological properties of these materials to date, it is clear that electronic and geometric surface properties can behave differently. This relationship might hold the key to developing a systematic picture of friction in layered compounds.

2.9 Charge Density Wave Systems

Charge density waves (CDWs) are periodic distortions in the charge density of a solid that are driven by Fermi-surface instabilities for wavevectors near $2k_F$. This rearrangement of charge results in lattice distortions (the Peierls distortion) and energy gas in the Fermi surface at sufficiently low temperatures (which may extend above room temperature). In addition, several CDW phases may exist for a single system at different temperatures. Because of their unique Fermi-surface behavior, low-dimensional materials include the layer-structure transition-metal dichalcogenides, such as $TaSe_2$ and TaS_2 (Coleman et al., 1985; Coleman et al., 1988; Thomson et al., 1988; Slough et al., 1989, 1990, 1991; Wang et al., 1990; Wu and Lieber, 1990a; Burk et al., 1991), as well as the quasi-one-dimensional transition-metal trichalogenides, such as $NbSe_3$ and o-TaS_3 (Gammie et al., 1989 a, b; Gammie et al., 1991), and metal-oxide bronzes (Nomura and Ichimura, 1990; Mercer et al., 1991; Rudd et al., 1991). In addition, defects and impurities can pin incommensurate CDW structures (Wu et al., 1988) and can perturb commensurate CDW superlattices (Wu and Lieber, 1990 b; Coleman et al., 1992).

Figure 82 shows an STM image from the first reported study of CDWs on layer-structure dichalcogenides (Coleman et al., 1985). Here, 1 T-TaS_2 was imaged at 77 K, and the lattice spacing observed is $\sim 3.5\ a_0$ (forming at $\sqrt{13} \times \sqrt{13}$ superlattice) with amplitudes of ~ 4 Å. At higher temperatures, transitions to incommensurate CDW phases occur, with a nearly-commensurate phase that can be seen at room temperature (Wu et al., 1988) since this phase exists in the temperature range 183–353 K.

Figure 83 shows a high-quality STM image of the $\sqrt{13} \times \sqrt{13}$ CDW structure of 1 T-TaS_2 obtained at 4.2 K; this CDW exists up to ~ 473 K (Slough et al., 1990). The accompanying line scan indicates that the corrugation due to atomic structure is $< 25\%$ of the CDW corrugation (Wang et al., 1990). While the measured CDW corrugations are in the range of ~ 2 Å, the actual CDW distortion in charge density is typically ≤ 0.2 Å. However, arguments on corrugation enhancement related to zone-boundary states (Tersoff, 1986), as discussed in the previous section, are applicable for CDW systems. At these low temperatures, the commensurate 1 T-$TaSe_2$ CDW system exhibits an energy gap, $2\Delta \approx 300$ meV, as shown in Fig. 83. It should be noted that a zero-bias anomaly has been found in a subject of such spectra, and these produce a sharp structure at $V_t = 0$, an overall conductance, a reduced barrier, and some enhancement in z-corrugation.

(a)

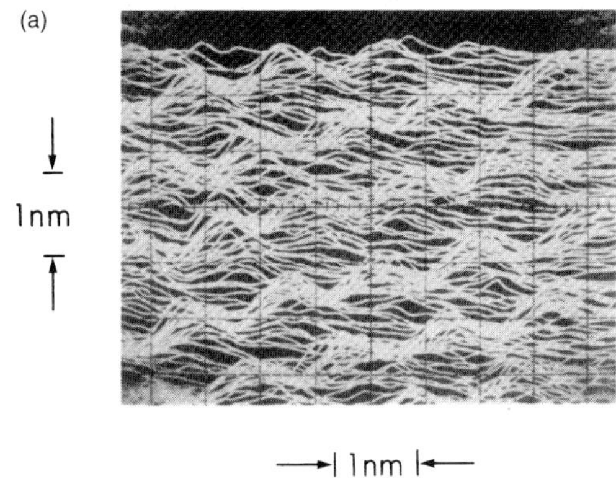

(b)

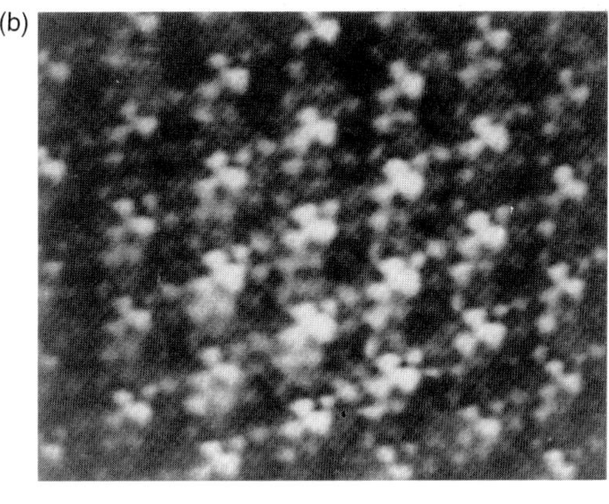

Figure 82. (a) Atom-resolved image charge density waves on the surface of the layer structure dichalcogenide 1 $T-TaS_2$. This early image of a transition metal dichalcogenide was acquired with $V_t = 50$ meV and $I_t = 2$ nA (Coleman et al., 1985). (b) The $\sqrt{13} \times \sqrt{13}$ superlattice rotated with respect to the underlying lattice by 13.9° (Coleman et al., 1988).

There has been controversy regarding the nearly-commensurate phase. It was predicted theoretically (Nakanishi and Shiba, 1977) that this phase would be incommensurate on the average but largely commensurate within a domain. Early studies showed no evidence for the predicted domain structure (Thomson et al., 1988). Domains were observed subsequently (Wu and Lieber, 1990 a), but other studies indicated that the structures were

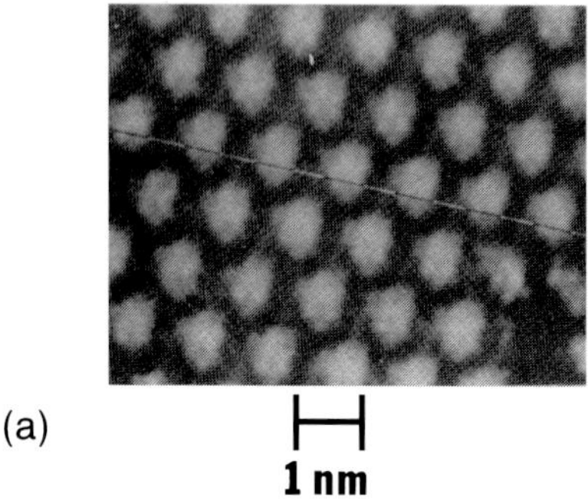

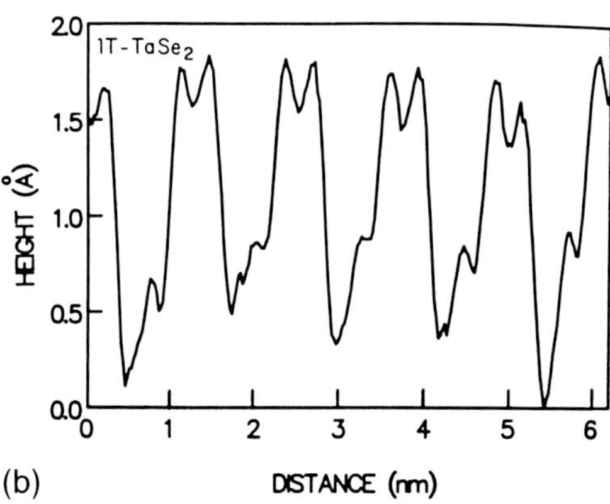

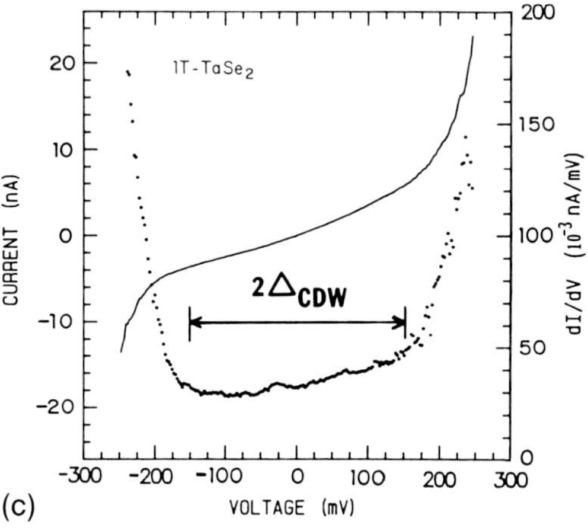

Figure 83. (a) A gray-scale image of $1T-TaSe_2$ obtained at 4.2 K illustrates both the CDW modulation ($3\ a_0$) and atomic modulation with (b) a corresponding STM scan profile. (c) I versus V and dI/dV versus V curves measured at 4.2 K showing the CDW gap (Wang et al., 1990).

continuously incommensurate (Slough et al., 1990). A detailed analysis of the Fourier-transformed images more recently demonstrated the shortfalls of visual interpretation of modulated superstructures, and confirmed that amplitude and phase domains do exist in the nearly-commensurate phase. The key conclusions of the study showed that the domains are not sharp; while the domain structures possess a real-space wavelength of ~ 72 Å, the boundaries comprise $\sim 50\%$ of the domain. In addition, there is extreme sensitivity to defects so that sample-to-sample variations can be significant.

Perturbations in a commensurate CDW phase due to defects and impurities have been investigated by STM for Ti-doped $1T$-$TaSe_2$ ($Ti_xTa_{1-x}Se_2$) (Wu and Lieber, 1990b). This material exhibits a commensurate CDW structure above room temperature, so measurements were performed under ambient conditions on freshly-cleaved surfaces. STM images of the surfaces of samples with $x = 0$, 0.02, 0.04, and 0.07, as shown in Fig. 84, depict the $\sqrt{13} \times \sqrt{13}$ superlattice. These images have regions of reduced CDW amplitude where the number of defects per CDW scales directly with the Ti concentration, and each defect apparently consists of several Ti centers. For $x \leq 0.4$, the phase of the CDW on either side of the defects does not vary, implying that an amplitude distortion is sufficient to relax the CDW around the defect. However, it is often the case that the image exhibits a coupled amplitude-phase distortion. In addition, in the regions around defects the CDW superlattice was sometimes found to exhibit a twin domain where there are $+13.9°$ and $-13.9°$ rotations with respect to the atomic lattice.

Increased disorder but large CDW corrugations with a reduction in the energy gap and lower conductance at zero bias in

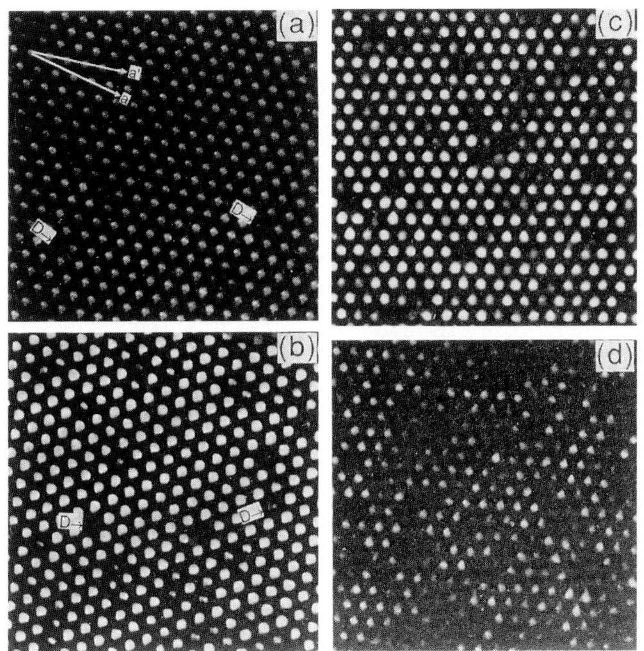

Figure 84. 220×220 Å^2 constant current images of (a) $TaSe_2$, (b) $Ti_{0.02}Ta_{0.98}Se_2$, (c) $Ti_{0.04}Ta_{0.96}Se_2$, and (d) $Ti_{0.07}Ta_{0.93}Se_2$ recorded with $I_t = 2$ nA and $V_t = 10$ mV. ***a*** and ***a'*** correspond to the lattice and CDW directions, respectively (Wu and Lieber, 1990b).

comparison to $1T$-TaS_2 (Coleman et al., 1992) has been observed for the $1T$-$Fe_{0.05}TaS_2$ system. Doping at this level was found to produce uniform spectroscopic effects over the entire crystal.

The quasi-one-dimensional transition metal trichalcogenides consists of chain structures, as illustrated in Fig. 85 for $NbSe_3$, with the *bc*-plane exposed for imaging (Gammie et al., 1989a, 1991). The Nb atoms of chain III physically protrude most, and this chain is staggered with respect to the other two by $b_0/2 = 1.7$ Å, as seen in Fig. 85. The strength of the Se–Se interaction is crucial to determine the CDW properties. The unit cells form sheets which are bonded together by van der Waals forces. Because of the weak inter-sheet and inter-chain bonding, these materials are fragile and can be difficult to image. There are two incommensurate CDW phases of $NbSe_3$ – one below 144 K, and one below 59 K. A CDW image obtained at 4 K (Slough et al., 1989), also shown in Fig. 85, gives the greatest superlattice modulation (~ 2 Å) with periodicity of $\sim 4\,\boldsymbol{b}_0 \times 2\,\boldsymbol{c}_0$. The CDW maxima comprise chains I and III, in contrast to images obtained at 77 K, above the low temperature transition, which favors chain II. Through these transitions, major alterations in the Fermi surface occur to produce the different structures as well as an approximately 30 meV CDW energy gap (Slough et al., 1989).

Another trichalcogenide, the orthorhombic phase of TaS_3 (o-TaS_3) was expected to exhibit similar structures to NbS_3 or monoclinic TaS_3 (m-TaS_3). However, the images were found to deviate significantly, as shown in Fig. 86, and no differentiation in the height between chains was observed (Gammie et al., 1989). This image rather corresponds to the structure of $ZrSe_3$, as shown. At 180 K, the CDWs begin to form, and, at 160 K, the CDW superlattice is fully formed, as shown in Fig. 86. The superlattice forms an angle of

(a)

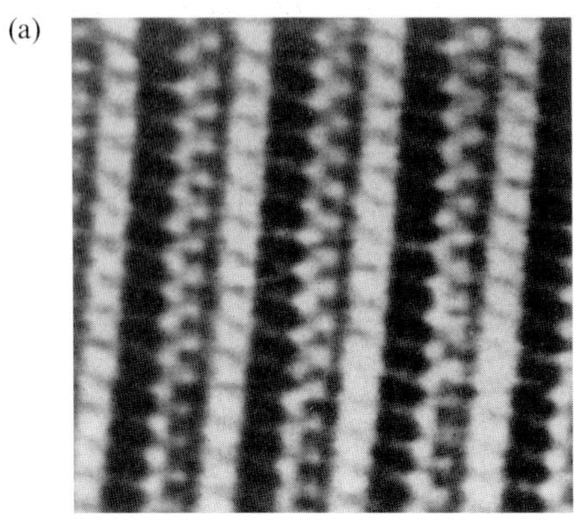

(b)

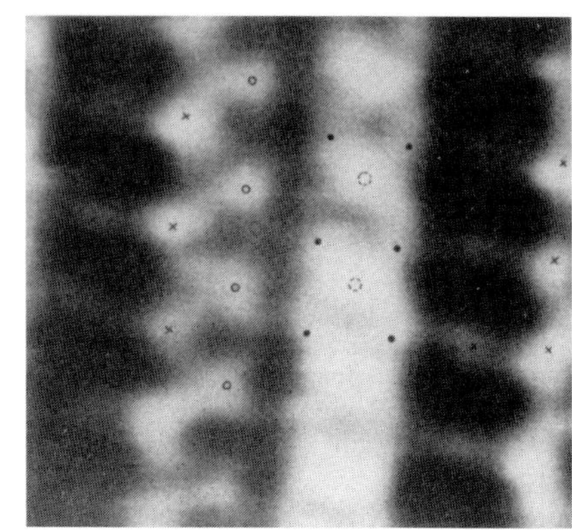

(c)

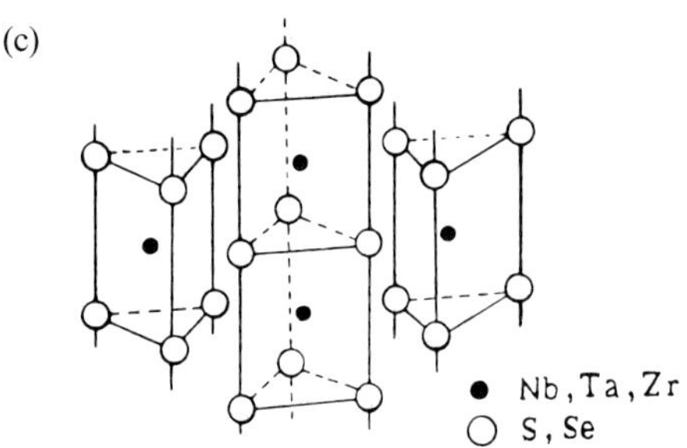

(d)

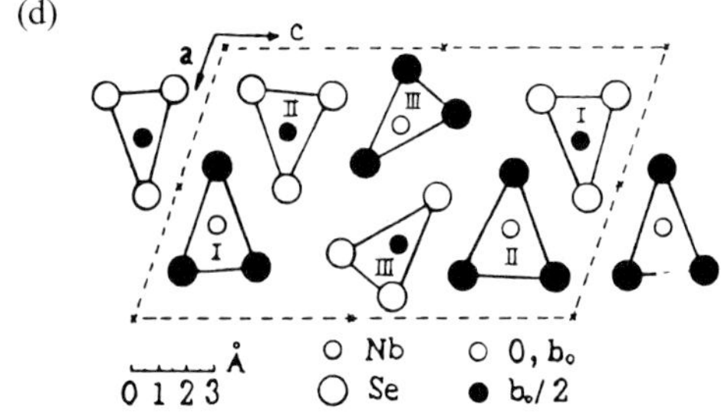

(e)

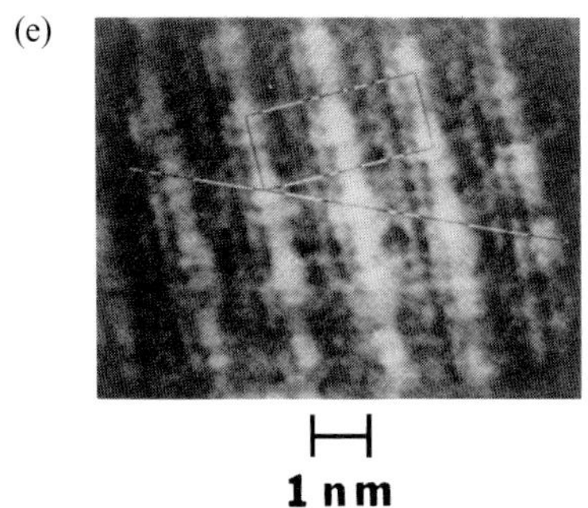

Figure 85. (a) 54×54 Å^2 and (b) close-up images of the b–c plane of $NbSe_3$ acquired at room temperature with the corresponding picture of the structure (c, d). The solid circles, crosses, and open circles correspond to chains, I, II, and III, respectively. Staggering of the chains, in correspondence with the model, is clearly seen (Gammie et al., 1989 a). (e) STM image of the CDW of $NbSe_3$ acquired at 4.2 K showing the CDW superlattice with unit cell dimensions of $\sim 4\,b_0 \times 2\,c_0$ (Slough et al., 1989).

$\sim 86°$ with the chain axis. In fact, two CDWs are seen to co-exist with the CDW maxima preferentially located on only two chains and not on the third. This indicates that the two CDWs, which might be expected to occur at different transition temperatures, are strongly-correlated forming a single Peierls transition, possibly due to the uniform interchain interaction and the effect of local coulombic repulsions associated with this structure (Gammie et al., 1989, 1991).

The quasi-one-dimensional molybdenum oxide bronzes ($A_xMO_zO_y$) are systems known to exhibit CDW behavior along the b-axis. Besides reports of the room temperature structure of the blue-bronzes $Rb_{0.3}MoO_3$ (Rudd et al., 1991) and $K_{0.3}MoO_3$ (Mercer et al., 1991), CDW transport at the surface of $K_{0.3}MoO_3$ at a

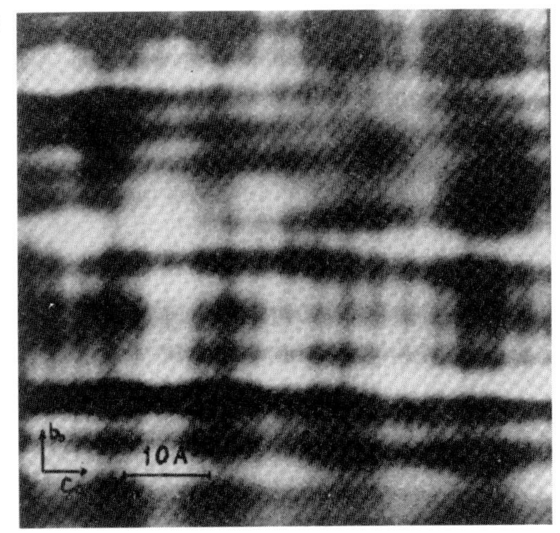

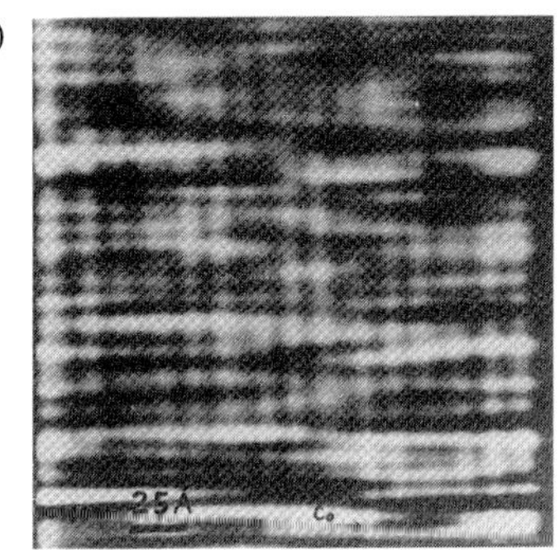

Figure 86. (a) 65 Å × 65 Å STM image of the trichalcogenide o-TaS_3 obtained at room temperature with the chain axis oriented horizontally and the scan direction along the c-axis. Atomic spacings of 3.3 Å and 2.5 Å are observed (Gammie et al., 1989 b). (b) 215 × 215 Å^2 image of the CDW superlattice of o-TaS_3 taken at 160 K with a horizontal wavelength of $4\,c_0$ and vertical modulations of $b_0/2$ (Gammie et al., 1989 b).

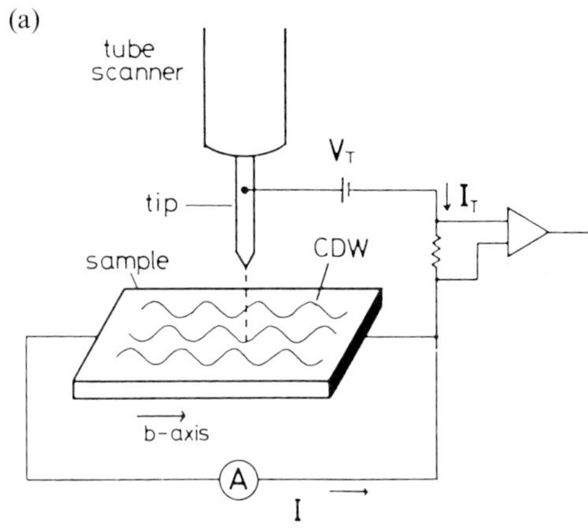

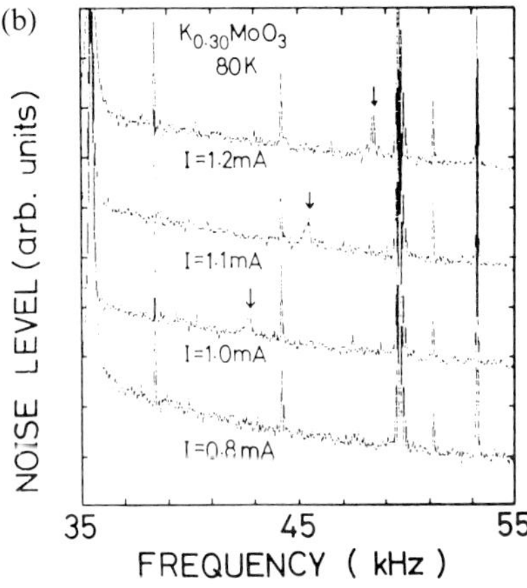

Figure 87. (a) Set-up to observe sliding charge density wave of $K_{0.3}MoO_3$. (b) Spectra of tunneling current under bias current with a fixed tunneling tip. Above the threshold a new peak appears, as shown by the arrow (Nomura and Ichimura, 1990).

temperature of 80 K was investigated (Nomura and Ichimura, 1990) by performing tunneling measurements for a fixed tip position and a parallel electric field with a magnitude above the de-pinning threshold applied. Fast-Fourier-transform-analyzed tunneling noise spectra exhibited sample current-dependent features related to the current carried by the electronic charge density modulation of the CDW, as shown in Fig. 87. At a CDW current of 1 mA, the sliding velocity was estimated at 5.7×10^{-3} cm/s, which is a factor of approximately ten times greater than in the bulk. In addition, Fig. 88 shows a measurement of the CDW gap of $K_{0.3}MoO_3$ at 77 K; the value of 2Δ obtained from these measurements was ~ 80 meV.

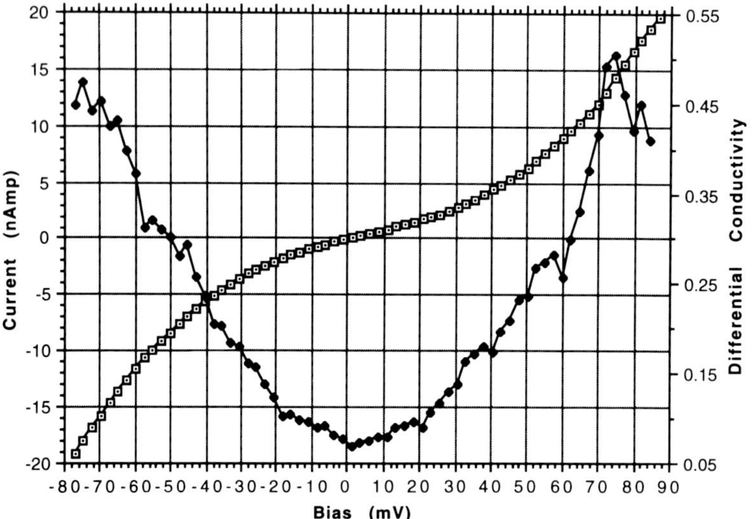

Figure 88. $I-V$ curve of $K_{0.3}MoO_3$ obtained at 77 K showing the CDW gap (Mercer et al., 1991).

2.10 Superconductors

The study of metal–insulator–superconductor and superconductor–insulator–superconductor tunneling, pioneered by Giaver (Giaver, 1960), has over the years become the basis for measuring excitations in the region of the superconducting energy gap. More recently, superconductor–insulator–superconductor junctions obtained by sample cleavage and reorganization in situ have been demonstrated (Moreland et al., 1987), and point-contact configurations have also been used (Hawley et al., 1986; Wilkins et al., 1990; Jiang et al., 1991). Both of these types of measurements provide valuable information on mechanically-stable junctions, but they lack the lateral spatial-resolving power of STM. Measuring the superconducting energy gap, Δ, with the STM follows directly from such studies and provides the ability to obtain spatial resolution. This sort of approach is particularly important in the study of new classes of high-T_c ceramics, which have drawn worldwide interest (Bednorz and Müller, 1986). The invention of STM fortuitously coincided with the discovery and development of high-T_c superconductors, for which the convergence of geometric and electronic structure is considered to be an important component in arriving at as yet unresolved superconductivity mechanisms. Measurements of superconducting samples demand instrumentation that can operate at variable cryogenic temperatures and with applied magnetic fields typically in a helium gas/liquid environment; one such instrument (Fein et al., 1987) has already been discussed in Sect. 2.3. On conventional superconductors, STM measurements of Δ at the surface can typically be directly compared to bulk values, but ceramic samples often exhibit deviations due to differences in stoichiometry between the surface and the bulk. The spatial-resolution of STM for topographic and spectroscopic images has also provided the unique ability to image the structure and motion of flux vortices on Type-II superconductors.

The first STM study of a conventional superconducting alloy (Elrod et al., 1984) was performed in vacuo on a thin-film Nb_3Sn sample with $T_c \approx 16$ K. Superconductor–insulator–normal metal tunneling was confirmed by the direct measurement of the barrier height, ϕ (~ 3.3 eV) at low temperatures. $I-V$ spectra obtained at $T \sim 10$ K detected a superconducting energy gap, Δ, of 3.7 meV, in agreement with the bulk value. In another study (Ramos et al., unpublished), $I-V$ measurements on Pb found an energy gap $\Delta = 1.25$ meV at 4.9 K and a spectral shape which fit well to BCS (Bardeen–Cooper–Schrieffer) theory.

The discovery of high-T_c superconductors has stimulated major efforts to obtain a comprehensive understanding of the chemical and physical aspects of these layered oxide materials. Resolving the mechanism for superconductivity in these materials experimentally requires the relation of structure, stoichiometry, and preparation to the transition temperature and energy gap. Modified surface stoichiometries and other materials problems can make the surface layers insulating, even though the bulk is superconducting. In addition, the coherence lengths are extremely small, of the order of a unit cell dimension. Thus, it is particularly important that local superconducting properties be correlated with structural images at the same location because of the difficulty of preparing these materials. Even with these problems, STM measurements have provided valuable new insights into these new materials by probing the local geometric and electronic structure (Kirtley et al., 1988 a).

Several structural studies have been performed at room temperature to assess the morphology and homogeneity of samples prepared by various methods. For example, atomic scale images of $Bi_2Sr_2CaCu_2O_{8+\delta}$, as shown in Fig. 89 (Kirk et al., 1988), obtained in vacuo show step heights in multiples of ~ 15 Å, indicating that cleavage occurs at whole unit cells or half unit cells. Imaging with a positive sample bias gave the best images suggesting that the unoccupied states images correspond to Bi (6p) orbitals of a metallic surface. Since only the top plane is accessible to STM, it was concluded that the Bi planes are the cleavage planes. Taking the bright spots forming squares aligned along *a-b* diagonals as Bi atoms, corrugations are observed approximately every 5 Å and striations are seen approximately every 26 Å along the *a*-axis. Detailed analysis of the corrugation amplitudes of atomic rows diagonal to this superstructure led to the conclusion that perpendicular *s*-wave lateral distortions result in a longitudinal displacement of ~ 0.5 Å along the *a*-axis and a buckling displacement out of the plane, the images of which may be complicated by electronic effects. Nevertheless, the superstructure occurs due to a lattice strain in the Bi–O plane, which results in a missing Bi row every ninth or tenth site. A subsequent analysis of this structure found that the corrugations for opposite polarities were in-phase (Shih et al., 1991). The results were inter-

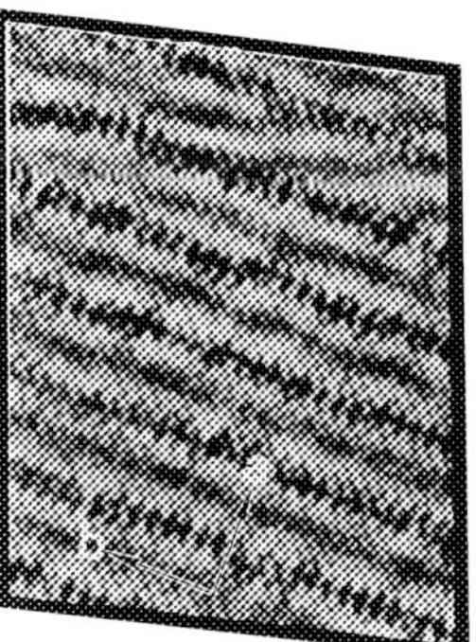

Figure 89. 150 Å × 200 Å image of $Bi_2Sr_2CaCu_2O_x$ acquired at room temperature with $V_t = -0.15$ V and $I_t = 0.2$ nA. In the proposed model, the square lattice of bright dots corresponds to the positions of Bi atoms. Note the superstructure that runs parallel to the *b*-axis (Kirk et al., 1988).

preted to be consistent with the imaging of BiO layers, although the non-metallic property of the surface presented the possibility that some effects come from the CuO layer beneath.

A study of a high-quality single crystal of $Bi_2Sr_2CaCu_2O_x$ (Liu et al., 1991) gave consistent spectroscopic features (Fig. 90) with $\Delta \approx 30$ meV in faceted field images with a ratio of zero-bias superconducting state conductivity to normal state conductivity of $\sim 1:12$. A linear conductance characteristic was observed for tunneling in the c direction in an unfaceted region. While analysis showed that the conductance spectra resemble that expected from BCS theory $[2\Delta/(k_B T_c) \approx 8.2]$, clear spectral broadening was noted here, as in other studies. In addition, a second peak at $\sim 3\Delta$ was observed and interpreted to originate from a structure in the quasiparticle density of states.

The nature of superconducting vortex cores in Type-II superconducting materials was mostly limited to theoretical study before the ability to image with low temperature, high spectroscopic resolution STMs became available. Recently however, observations of the motion of unpinned flux vortices (Dittrich and Heiden, 1988) and images of the Abrikosov flux lattice (Hess et al., 1989; Hess et al., 1990; Renner et al., 1991) have been reported.

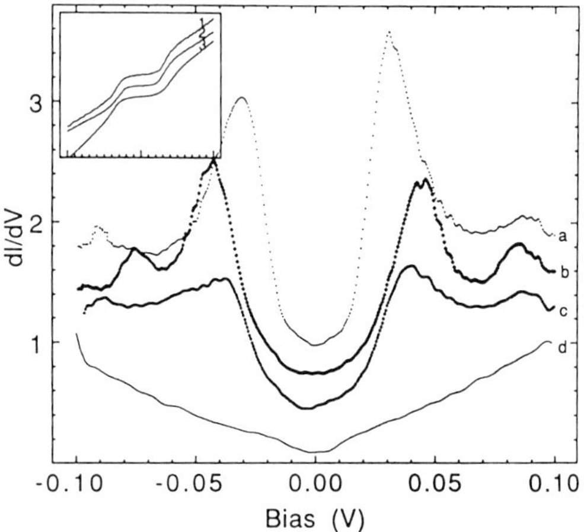

Figure 90. (a–c) Normalized plots of dI/dV versus V on the center of a $Bi_2Sr_2CaCu_2O_x$ facet showing the superconducting energy gap and ± 90 meV peaks. (d) A plot for tunneling in the c-direction (Lie et al., 1991).

The motion of superconducting flux vortices at the surfaces of Type-II superconductors was detected in an experimental arrangement to detect quasiparticle tunneling between a fixed Nb tip and a Nb single crystal surface in a magnetic field (Dittrich and Heiden, 1988). The Nb sample was placed in a mixed state by the application of a magnetic field while a current was passed through the sample to drive a vortex current. Random modulation of the gap voltage with time (vortex motion "noise"), observed on an oscilloscope, signified the passing of superconducting vortices with normal cores beneath the tip. To support the conclusion that gap fluctuations originated from vortex motion, the "noise" was studied as a function of applied magnetic field and sample current. In particular, above a critical field H_{c2} noise fluctuations were absent; likewise, fluctuations were not observed below a critical sample current due to vortex pinning.

The spatial dependence of the quasiparticle energy spectrum for flux vortices was first observed on $2H$-$NbSe_2$ (Hess et al., 1989), a layered material with atomic spacing: $a_0 = 3.5$ Å; $T_c = 7.2$ K; upper critical field, $H_{c2} = 5$ T; and coherence length, $\xi = 77$ Å. The charge-density-wave transition (periodicity $\sim 3\,a_0$) was confirmed at 33 K. The observed superconducting gap was 1.1 meV, in good agreement with other measurements.

Imaging the Abrikosov flux lattice can be accomplished at low temperatures with bias voltages of the order of mV, because the electronic density of states is modulated by the vortices which exist in the presence of an applied magnetic field. At these biases, the superconducting regions exhibit low conductance (I/V), while the normal cores have a higher density of states. The low temperature also suppresses Fermi level smearing, which could easily be of the order of the gap width. Figure 91 shows the flux lattice (vortex spacing of ~479 Å) obtained by the application of a ~1 T magnetic field. Spatially-resolved spectroscopic measurements across the flux lattice found an enhancement in the DOS at the center of the vortex rather than a flat, featureless normal metal spectrum as expected (Fig. 91). Farther away from center, but still in the core, the DOS flattened, and the boundary regions exhibited a clear superconducting energy gap. Subsequent analysis of these observations concluded that the zero-bias enhancement of the DOS on the center originates from the lowest quasiparticle bound state (Gygi and Schluter, 1990). Exact agreement with the observed spectral shape required a broadening mechanism, such as defects or finite temperature. Further details of the flux lattice, including reduction of the band gap due to the superfluid velocity far away from the core, distortion of the vortex images due to the interaction of the crystal band structure with the CDW gap and the effect of vortex–vortex interactions were explored in subsequent work (Hess et al., 1990). In addition, the $Nb_{1-x}Ta_xSe_2$ system was studied as a function of x to investigate the role of impurities (Ta) and disorder on the behavior of the zero-bias conductance in the vortex core region (Renner et al., 1991). As seen in Fig. 91, a peak in the zero-bias conductance on the center was observed as expected for $x = 0$, but this enhancement was removed for $x = 0.20$. A pair-breaking

(A)

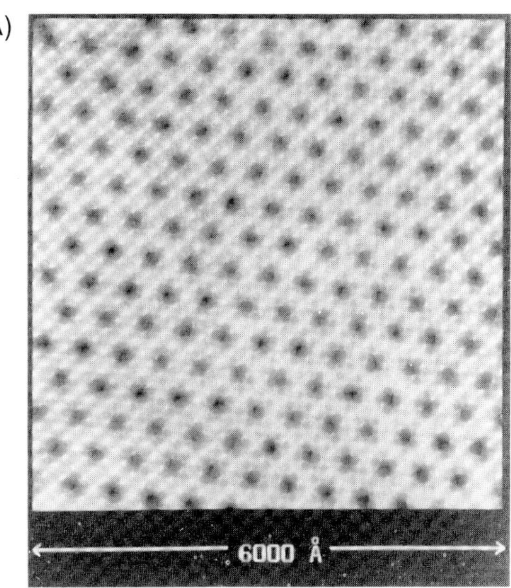

(B)

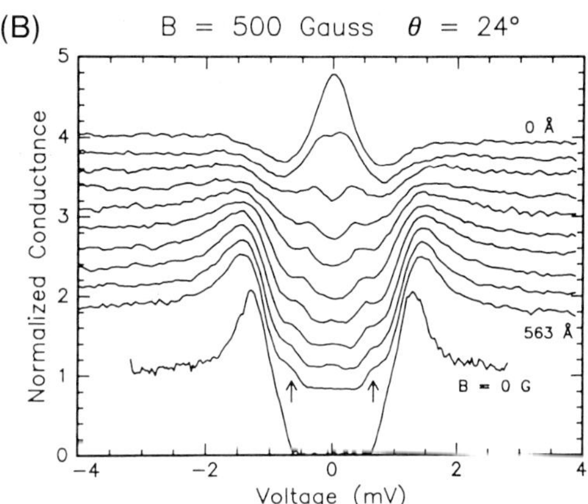

(C)

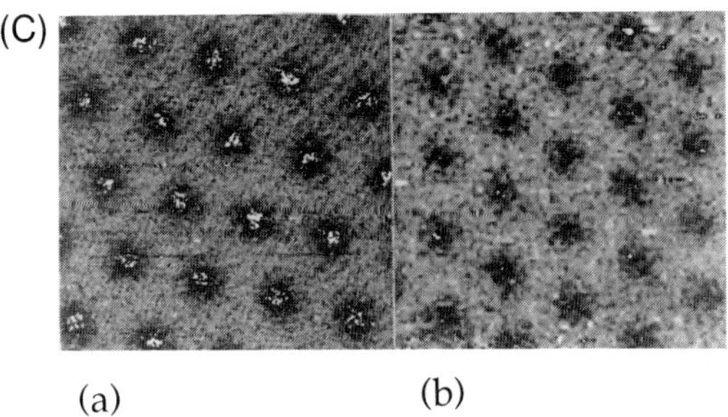

(a) (b)

Figure 91. (A) dI/dV image of the Abrikosov flux lattice on $NbSe_2$ for a ~1 T magnetic field at 1.8 K (Hess et al., 1989). (B) dI/dV versus V for various positions moving through a vortex core; at the center (0 Å) there is a zero-bias enhancement. The lower curve is for zero applied field $B = 0$ (Hess et al., 1990). (C) 4000 Å × 4000 Å STS images of the vortex lattice at 1.3 K and 0.3 T for (a) pure $NbSe_2$ and (b) $Nb_{0.8}Ta_{0.2}Se_2$ where the zero-bias conductance in the center of the core is quenched (Renner et al., 1991).

model to insert the effect of inelastic scattering on the DOS was put forward to explain this behavior.

2.11 Molecular Films, Adsorbates, and Surface Chemistry

Imaging atomic and molecular structures at surfaces using the combined structural and spectroscopic capabilities of the STM provides a means to probe the chemistry and bonding of monolayer, as well as multilayer films. Adsorption and chemical reactivity at solid surfaces are local phenomena that require knowledge of the local structure, and the STM technique can provide this information. Beyond traditional (and non-traditional) surface chemistry performed in ultrahigh vacuum environments, the added capability to image molecular crystals, polymer films, and biological molecules in air or liquids opens a vast range of possibilities, including the realization of nanoscale manipulation of a surface.

2.11.1 Molecular Imaging

The first demonstration of the ability of STM to obtain high resolution images of organic molecules in air was reported for the conducting molecular crystal tetrathiafulvalene tetracyanoquinodimethane (TTF-TCNQ) (Sleator and Tycko, 1988). Images revealed raised rows of ball-like structures with diameters of ~3 Å surrounded by two rows of smaller ball-like structures, as shown in Fig. 92. Calculations led to the assignment of this triplet structure to a single TCNQ molecule, with the smaller balls centered over N atoms and the center ball an image of the electron density of the center ring. STM studies of liquid crystals followed soon afterwards demonstrating, for example, one- and two-dimensional ordering of individual molecules of the liquid

(A)

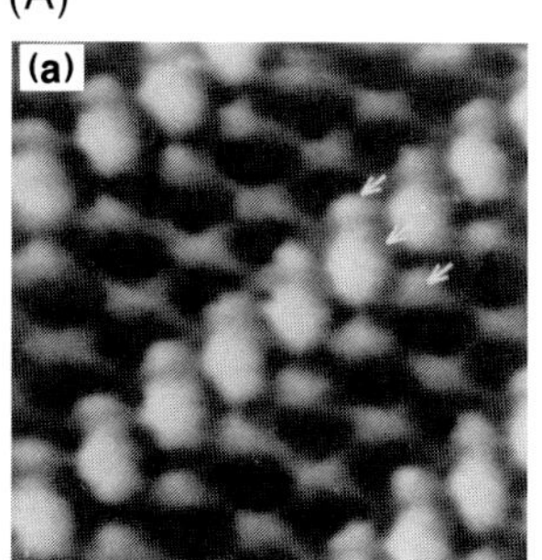

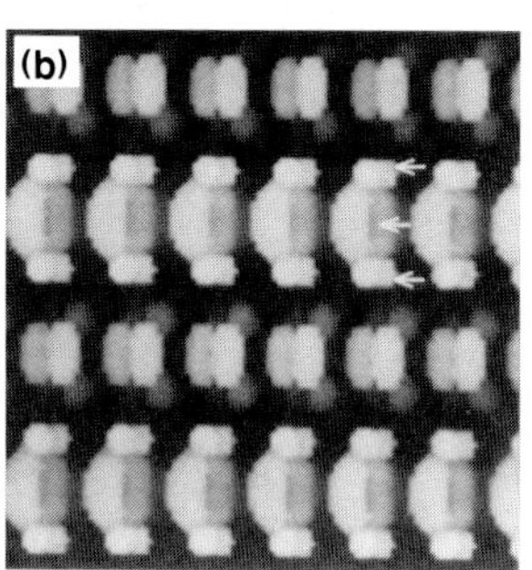

(B)

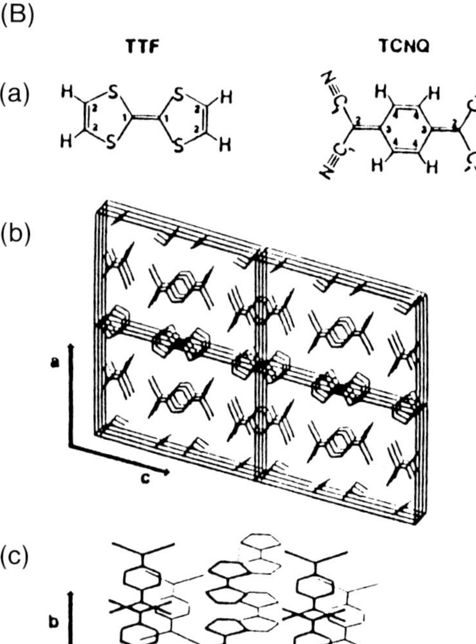

Figure 92. (A) 25 × 25 Å^2 STM image of TTF-TCNQ in air (a) and a simulated image (b). (B) Molecular structure of TTF-TCNQ as observed in the image (Sleator and Tycko, 1988).

crystal deposited on a graphite single crystal (Foster and Frommer, 1988) and the identification of individual functional groups of these species (McMaster et al., 1990).

While bias-dependent imaging provides local electron geometric and spectroscopic information, studies where Langmuir–Blodgett (L–B) films and organic multilayers are imaged indicate that the tip may penetrate the adsorbed species without damage, but with possible distortions related to mechanical, chemical, or electrostatic interactions between the tip and the sample (Fuchs et al., 1990). Furthermore, the issue of electron transport through an insulating film must be considered in order to prevent charging, and to obtain contrast. One possibility is that the tunnel barrier of the insulating film is smaller than the adjacent vacuum in gap in order to measure topography in the constant current mode. Another possibility is that local electron states are accessible at the surface for tunneling and that these states conduct electrons to the substrate faster than the arrival rate of tunneling electrons (Coombs et al., 1988). This requires high mobility in the substrate, which might occur by tunneling through overlapping molecular orbitals, bulk hopping-type conduction, Schottky emission into the conduction band of the insulator, or field-induced ionization in the insulating film (Dietz and Herrmann, 1990).

Following the first report of imaging of thin film Langmuir–Blodgett deposited polymers (Albrecht et al., 1988), STM studies of the technologically-important polyimide poly-[*N*,*N*′-bis(phenoxyphenyl)pyromellitimide] deposited on an HOPG surface by Langmuir–Blodgett techniques (Sotobayashi, 1990; Fujiwara et al., 1991), or on gold films by vapor deposition and thermal imidization (Grunze et al., 1988),

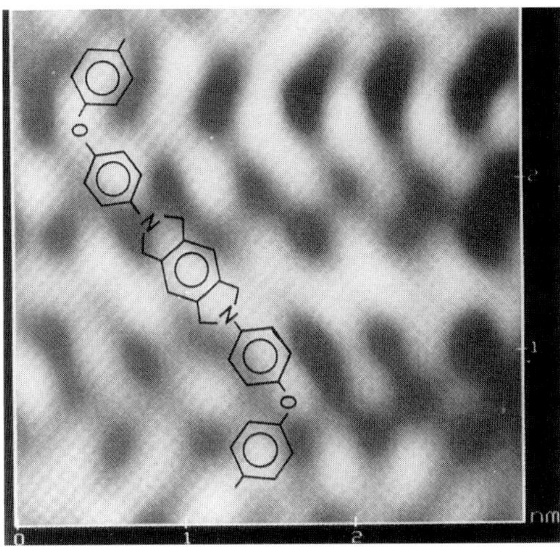

Figure 93. STM image of a polyimide thin film deposited on graphite by Langmuir–Blodgett techniques (Fujiwara et al., 1991).

were reported. Images, as shown in Fig. 93, indicate the packing structure on the substrate. In particular, the individual polyimide chains exhibit a zigzag structure with bending angles of 110–120° at the ether-bonded O atoms and an O–O distance of 15–20 Å. The chains appear to be aligned parallel to the deposition direction of the Langmuir–Blodgett film. The chain–chain distance was found to be ~5 Å, which suggests a tilting of the entire molecule to avoid steric hindrance effects.

2.11.2 Adsorption and Surface Chemistry

From the surface science point of view, the formation of a chemical bond and the occurrence of a chemical reaction at a surface is a local phenomenon occurring at one or more specific atomic sites. Studies of electronic structure and chemistry at surfaces performed by techniques that average over the surface can provide much information. However, since adsorption species are typically associated with specific atomic sites, and since molecular dissociation and catalytic reactions often occur preferentially on specific surface planes, at steps, or at defects, an atomic-scale spectroscopic probe such as an STM/STS is the most appropriate. The difference between STM images of the clean surface and the adsorbate-covered surface can elucidate the local chemical interaction due to adsorbate-induced changes in the local density of states of the valence electrons that are directly involved in the bonding interaction. When combined with results of other surface techniques, STM and STS make it possible to assess site-specific electronic interactions related to bonding and chemical reactivity, as well as the spatial distribution of reactions on solid surfaces. We have already discussed the role of adsorbates in metal reconstructions. Here we focus on imaging the adsorbed species themselves and deriving geometric, electronic, and chemical information at metal and semiconductor surfaces.

Semiconductor Surfaces

The covalent bonding nature of semiconductors uniquely determines adsorption and surface reactivity. For example, their half-filled dangling bonds extending out into the vacuum are typically very active reaction sites. The formation of new structures and the passivation of the dangling bonds may involve bond-breaking, which is kinetically-limited at room temperature, but direct reaction at higher temperatures or annealing after room temperature deposition can promote such processes. We consider important examples where STM has been used to derive information on Si surface chemistry: (1) molecular dissociation and bonding (Hamers et al., 1987; Wolkow and Avouris, 1988); (2) the incorporation of subsurface boron and the resulting chemistry of that system (Lyo et al., 1989); (3) etching of Si adatoms by H or Cl (Villarrubia and Boland, 1989; Boland, 1990, 1991 a), (4) oxidation reactions at room temperature (Avouris and Lyo, 1992) and at elevated temperatures (Feltz et al., 1992); and (5) epitaxial growth of CaF_2 insulating films (Avouris and Wolkow, 1989 b). In addition, we consider the long-range effect of adsorbate interaction with a GaAs surface due to band bending effects (Stroscio et al., 1987).

The first applications of STM to observe the atomic-site dependence of reactions on semiconductor surfaces focused on the adsorption of ammonia (NH_3) by Si(100)-(2 × 1) (Hamers et al., 1987) and Si(111)-(7 × 7) (Wolkow and Avouris, 1988) surfaces.

In the case of the Si(100)-2 × 1 surface, Fig. 94 shows an occupied state image of the clean Si(100)-2 × 1 surface (see also Fig. 22). Dosing the surface with NH_3 was found to retain the 2 × 1 symmetry, while the electron density between the dimers was reduced with the tunneling enhanced at the dimer ends (c). Measuring spectra before and after exposure shows that the bonding and anti-bonding π states are removed. An adsorbate-induced shoulder forms ~ 1 eV above E_F which is attributed to a Si–H antibonding state. With reference to related studies, the picture that emerges is that the NH_3 dissociates and a monohydride H phase is formed at saturation coverage. It has been concluded from ion scattering studies that N occupies subsurface sites, so N is unlikely to appear in the images for this reason, as well as the fact that the bonding state is ≥ 4 eV, which is beyond the range of STS.

It was previously shown that the various local configurations of surface atoms within a 7 × 7 unit cell produce site-specific electronic structures; charge transfer be-

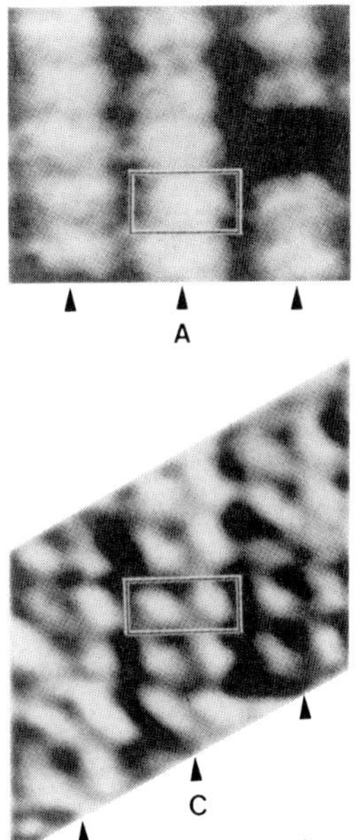

Figure 94. STM images of occupied states on a clean (upper) and NH_3-dosed (lower) Si(100)-2 × 1 surface (measured with $V_t = 2$ eV) showing the reduction in surface state density between the atoms due to the removal of bonding π levels (Hamers et al., 1987).

tween the different types of Si atoms and surface strain are expected. On extending the NH_3–Si reaction to the Si(111)-7 × 7 surface, marked differences in local chemical activity might be expected. Previously, an electron energy loss study of the surface vibrational structure (Tanaka et al., 1987) showed that NH_3 dissociatively adsorbs at room temperature leaving H–Si + NH_2–Si species; corresponding changes were also observed by photoelectron spectroscopy. Figure 95 shows an unoccupied state image of the clean Si(111)-(7 × 7) surface ($V_t = -0.8$ eV). The twelve adatoms of the unit cell are clearly seen, and spectra for the rest-atom, corner adatom, and center adatom sites are presented. After partial reaction with NH_3, the image of the surface adatoms appears distinctly different; with $V_t = -0.8$ eV it appears that some adatoms have been removed. However, the adatoms are observed at a higher bias ($V_t = -3$ eV), so the disappearance of the

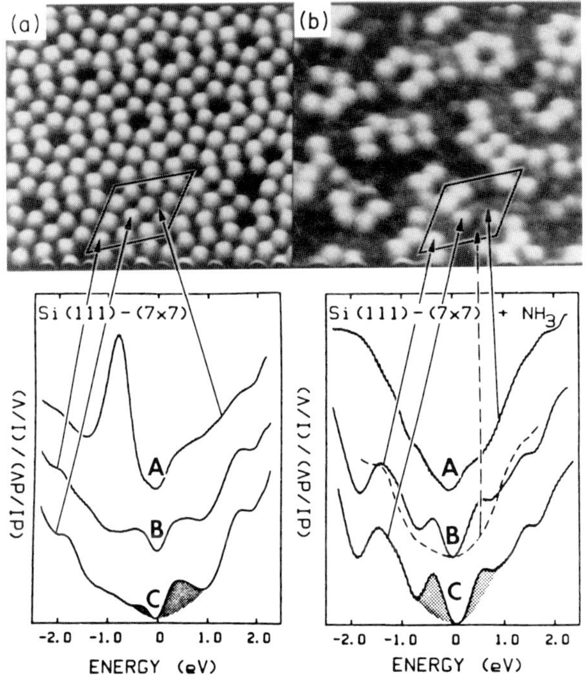

Figure 95. Upper panel: unoccupied states of (a) clean ($V_t = -0.8$ eV), and (b) NH_3 partially reacted surfaces ($V_t = -0.8$ eV). Lower panel: for the clean surface, tunneling spectra over a rest atom site (A), a corner adatom site (B), and a center adatom site (C); the shaded regions depict the differences from bulk Si. For the NH_3-exposed surface, tunneling spectra over a reacted rest atom site (A), a reacted corner atom site (B, dashed), an unreacted corner atom site (B, solid), and an unreacted center atom site (C) (Avouris and Wolkow, 1989a).

adatoms at $V_t = -0.8$ eV is not due to their physical removal but is related to an adsorbate-induced change in the state density at these positions. Local tunneling spectroscopy supports this assertion. Under these conditions, the rest atom state as well as the surface states related to the rest adatoms and corner adatoms are removed. It appears that the rest atoms are more reactive than either type of adatom, partly due to the fact that the adatom dangling bond state is more delocalized than the rest atom state. The center adatoms are more reactive than the corner adatoms, considering the relative number of reacted adatoms. It can be deduced from the spectra that, upon reaction, both types of adatom sites become similar, with the occupied states having grown in intensity. Reaction-induced charge transfer from the rest atom to the adatom was proposed to account for this process, and a new state is formed in (B) at ~ -1.5 eV. A picture was advanced in which the rest atom dangling bond state bonds to either H or NH_2; these products can also be scavenged by the other dangling bonds or an Si–Si bond can be broken. Since there are two center adatom neighbors for each rest atom versus one corner adatom neighbor, a geometric origin for the increased reactivity of the center adatoms may be suggested.

The geometric, chemical, and electronic properties of Si(111)-7 × 7 surfaces exposed to boron, followed by annealing, are particularly interesting cases of chemical modification of surface electronic properties. B is a common semiconductor dopant, and a set of studies (Bedrossian et al., 1989; Headrick et al., 1989; Lyo and Avouris, 1989; Lyo et al., 1989; Thibaudau et al., 1989; Avouris et al., 1990) began with what is referred to as a "δ-doped" interface, since structurally the B resides in a single subsurface layer. We note that the electrostatic effects of the layer extend over a significantly longer range.

Group-III atoms such as Al form covalent bonds with the T_4 dangling-bond sites of the Si(111) surface to allow pairing of all the valence electrons; a $\sqrt{3} \times \sqrt{3}$ structure is the result. B behaves rather differently in that the on-top T_4 site is unstable due to the stress that would come from shorter bonds than the other Group-III atoms. It was found that B adopts a substitutional site beneath the T_4 site (Headrick et al., 1989); this bonding configuration requires a significant charge transfer that modifies the surface electronic structure. One method to prepare the substitutionally-doped surface is to expose the clean Si(111)-7 × 7 surface to decaborane ($B_{10}H_{14}$) and anneal at 1000 °C to drive off the H.

Before being incorporated into the subsurface, B can be observed on top of the surface. A partial surface layer of B obtained by a decaborane exposure of 0.4×10^{-6} torr s and annealing produced the image of Fig. 96 A, acquired at $V_t = -2$ V. The structure is $\sqrt{3} \times \sqrt{3}$, but a considerable number of dark areas are apparent. These dark areas are assigned to the B–T_4 which are lower and have more contracted wavefunctions; the bright atoms correspond to Si–T_4 adatoms. Since bonds must be broken, there are kinetic limitations to the formation of this structure, but annealing a sample exposed to 1×10^{-6} torr s of decaborane at 1000 °C produced a high-quality $\sqrt{3} \times \sqrt{3}$ structure, as shown in Fig. 96 B, that exhibits a band gap of 2 eV, as determined by STS. The STM/STS data support the substitutional B model. The surface is very different chemically in that it is very unreactive because of the removal of dangling bonds (Avouris et al., 1990).

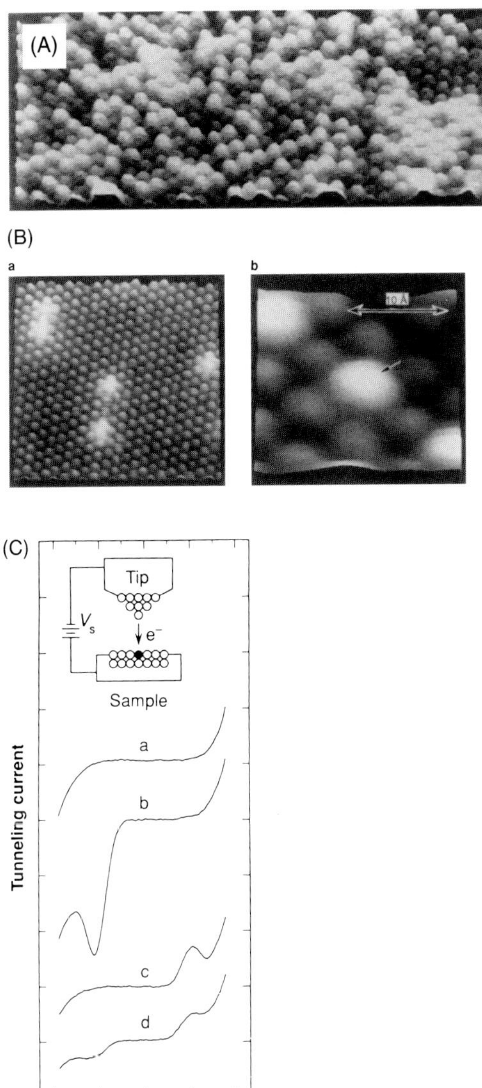

Figure 96. (A) Unoccupied state image ($V_t = -2$ V) of the Si(111) ($\sqrt{3} \times \sqrt{3}$) $R30°$-B surface formed by exposure to 0.4×10^{-6} torr s decaborane and annealing at 800 °C. (B): (a) Unoccupied state STM image ($V_t = -2$ V) of the Si(111) ($\sqrt{3} \times \sqrt{3}$) $R30°$-B surface formed by exposure to 1×10^{-6} torr s decaborane and annealing at 1000 °C with (b) a higher magnification image of a localized defect which exhibits a unique spectroscopic characteristic. (C) (a) $I-V$ spectrum from ideal terraces on the Si(111) ($\sqrt{3} \times \sqrt{3}$) $R30°$-B surface, (b), (c) spectra near defect sites exhibiting negative differential resistance (Lyo and Avouris, 1989; Avouris et al., 1990).

The image in Fig. 96 B indicates that the $\sqrt{3} \times \sqrt{3}$-B surface contains both atomic-scale and extended defects. These are the bias for perhaps the most interesting aspect of this surface – its tunnel-diode $I-V$ characteristics. In particular, the defects on the $\sqrt{3} \times \sqrt{3}$-B surface were found to exhibit negative differential resistance (NDR) (Lyo and Avouris, 1989) similar to the $I-V$ curves for Esaki diodes and quantum-well structures. $I-V$ spectra are depicted in Fig. 96 C for various points on the surface. (a) shows the ~2 eV band gap of the substitutional-doped B structure. (b)–(d) were obtained atop defect sites, and NDR appears where $dI/dV < 0$. Observation of this effect appears to depend on the presence of localized states (≤ 10 Å in spatial extent), possibly a vacancy involving the absence of an Si atom over a subsurface B atom. By simulations, it was demonstrated that the presence of localized states on the tip, possibly due to the capture, of an Si atom on the tip, as well as the sample, could account for spectra showing NDR. This sort of localized chemistry and electronic structure could be a precursor to the fabrication and operation of actual nanoscale devices.

Hydrogen is known to form mono-, di-, and tri-hydride species on Si(111)-7 × 7 with room temperature saturation H coverage greater than that needed to passivate the dangling bonds of the (7 × 7) surface. The presence of Si–H_x species is consistent with the possibility that atomic H etches the surface. An STM study of this surface (Boland, 1991 a) found a marked rearrangement of the top adatom layer for room temperature saturation coverage. The rest atom layer beneath remains intact with isolated adatoms and adatom clusters left above, as seen in the empty-state topograph of this surface in Fig. 97. A band gap of almost 2 eV was observed for the isolated atoms and the rest layer, signi-

(A)

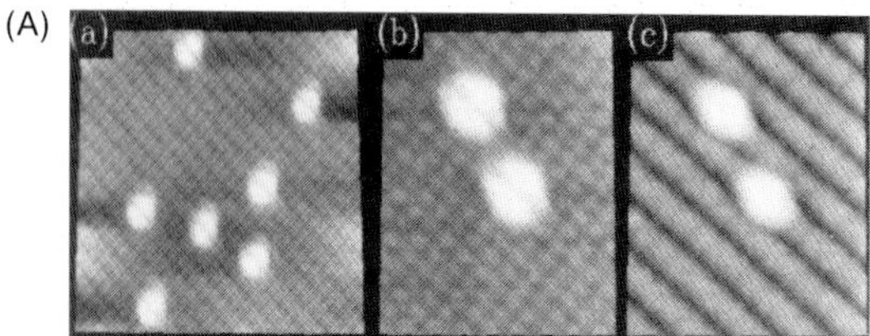

(B)

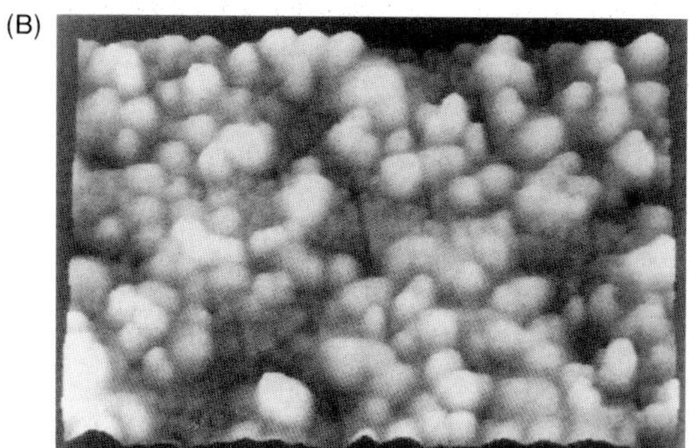

(C)

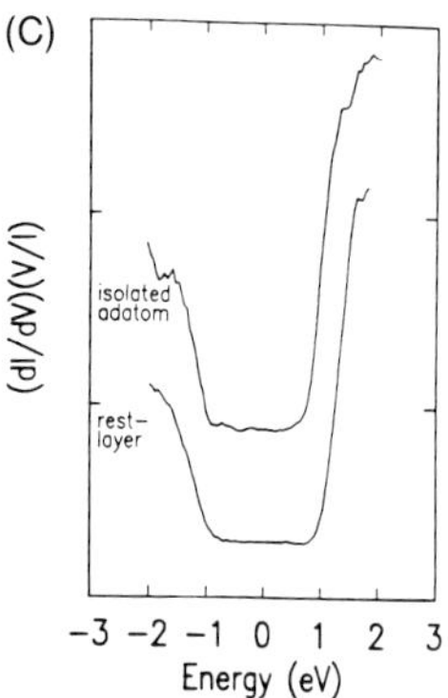

Figure 97. (A): (a) 125×130 Å^2 STM image of the Si(100)-2 × 1:H monohydride surface after 10 s annealing at 690 K; (b) 46×62 Å^2 STM occupied state image of the surface with $V_t = -1.7$ V; (c) 46×62 Å^2 STM unoccupied state image of the surface with $V_t = 2.0$ V (Boland, 1991 b). (B) 230×230 Å^2 STM occupied state image of atomic hydrogen saturated Si(111)-7 × 7 at room temperature, recorded with $V_t = -2$ V. (C) STS spectra of H-saturated Si(111)-7 × 7 surface over the rest layer and over an isolated adatom (Boland, 1991 a).

fying that the dangling bonds are saturated. A determination of the registry of the isolated adatoms indicated that the Si adatoms sit on a site atop of the rest layer, which is consistent with an unstrained tri-hydride species. With the assumption that the adatom strain energy, E, goes as $E_{\text{monohydride}} > E_{\text{dihydride}} > E_{\text{trihydride}}$, the kinetics of the reaction favor the formation of a trihydride so that the adatoms must move from a T_4 bonding configuration to an atop site. Annealing was found to further reduce strain in the rest layer by removing dimer bonds at the stacking fault with a reduction in corner holes directed towards the formation of a 1 × 1 structure.

Strong evidence for local chemical interactions at clean surfaces which determine the initial stages of reactivity has already been presented. Understanding surface chemistry at this level can provide insight into subsequent states of the reaction process. Oxidation of Si is an important process in microelectronics technology. Starting with the clean Si(111)-7 × 7 surface at room temperature, it has been suggested that the initial stage of oxidation involves a molecular precursor – an O_2^- species (Höfer et al., 1989). STM/STS studies (Avouris et al., 1991; Avouris and Lyo, 1992) have produced a model that considers the dissociation of the precursor species and the reaction of atomic oxygen with the surface. Figure 98 shows an unoccupied state image ($V_t = -2$ V) of a region of Si(111)-7 × 7 surface after 0.2×10^{-6} torr s exposure to O_2 at room temperature. Some sites appear brighter and some darker than the unreacted Si sites. Previous STM studies (Pelz and Koch, 1990) performed after considerably larger O_2 exposures found the surface already roughened, so that the detail of Fig. 98 was impossible to achieve. Altered sites (brighter or darker) appear more often on the faulted halves; these features are more prevalent at corner-adatom sites than at center adatom sites. The bright regions and the dark regions are not correlated in their growth suggesting separate reaction path-

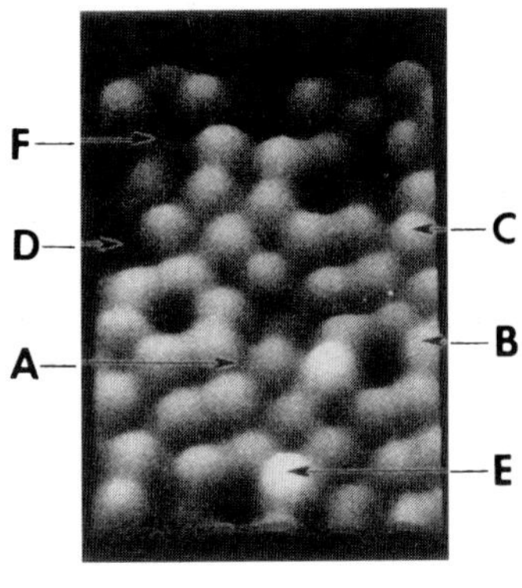

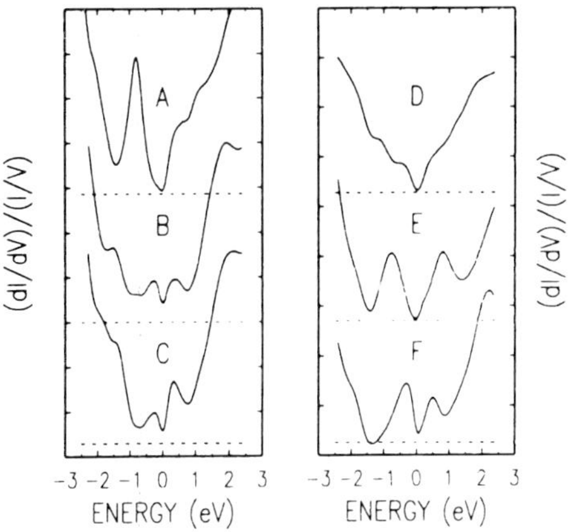

Figure 98. Upper panel: STM image of O_2-exposed Si(111)-7 × 7 at room temperature, showing the appearance of bright and dark features coexisting with the unreacted Si adatoms. The faulted and unfaulted halves are indicated. Lower panel: STS spectra obtained over six labeled sites subsequent to O_2 exposure at room temperature (Avouris et al., 1991).

ways, and dark regions increase steadily with increasing O_2 exposure, indicating this to be the main reaction pathway.

By comparing STS with theoretical calculations, it was concluded that a bright sphere in the unoccupied state image is Si adatoms that are left with an empty dangling bond due to interaction with a neighboring oxygen atom. In addition to these stable bright features, time-dependent bright features may be due to the populated low-lying levels of a long-lived (~ 10 min) molecular precursor at room temperature. Features which appear dark for both biases are due to the saturation of an Si dangling bond by oxygen removing low-lying surface states, and are therefore the most likely oxidation product of the reaction. The increased electron density near E_F of the faulted half apparently promotes molecular dissociation through interaction with the $2\pi^*$ affinity level of O_2, resulting in the formation of a negative ion. Once the molecule is dissociated, the corner-adatom sites are the favored site due to increased electron density in their vicinity, so providing an explanation for site-dependent reactivity in the oxidation process.

In contrast to the initial stages of oxidation at room temperature, oxygen etching of Si surfaces for microelectronics processing applications occurs at high temperatures with SiO(g) as the reaction product. This etching process was recently imaged in real time using an STM operating at ~ 950 K in an O_2 atmosphere at a pressure of ~ 5×10^{-6} Pa (Feltz et al., 1992). A sequence of images (Fig. 99) each separated in time by ~ 150 s taken under these conditions shows the evolution of the etching process. Part (a) shows three atomically-flat terraces separated by linear step edges with a few kinks. Scanning this area after ~ 150 s shows a loss of material at the steps with general step roughening, as seen in (b). This is due to the fact that, at this temperature, O_2 dissociates on the surface, preferentially reacting with and removing Si atoms at step edges, so resulting in a vacancy diffusion process. In (c), only extended fingers of the original terraces remain. The end points of the fingers are typically pinned at the position of kink sites on the original steps. Once the topmost Si layer is removed, the next layer exhibits an imperfect 7 × 7 structure with a high density of defects. In (d) the bases of

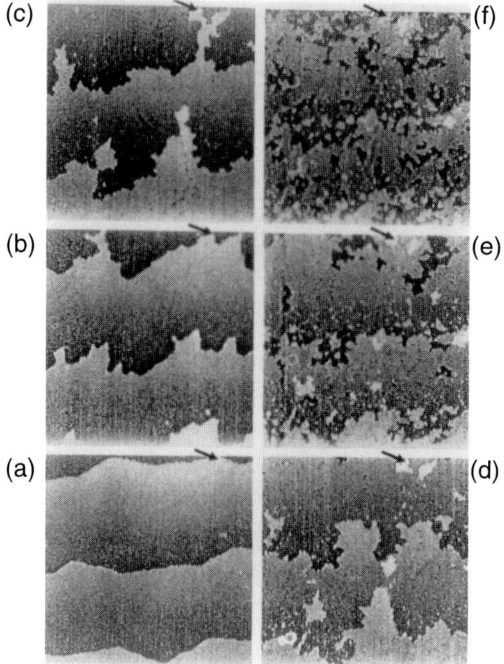

Figure 99. Sequence (a ⇒ f) of 2000 × 2000 Å^2 STM images of the Si(111)-7 × 7 surface at 950 K during exposure to 5×10^{-6} Pa oxygen. The time between images was ~15 s. (Feltz et al., 1992).

the fingers are etched and small islands and small holes begin to form, as seen in (e), with new smoother step edges appearing at the positions of the original steps. The defects act as nucleation centers for continued etching. As seen in (f), an increasing number of monolayer-deep pits is seen whereby an atomically-rough surface remains with three layers simultaneously exposed. These reactions are limited by kinetics, and the resulting structures are thermodynamically unstable, as demonstrated by surface smoothing upon annealing at 1050 K, once the O_2 atmosphere was removed.

Ultrathin insulating films on semiconductor surfaces are important in the microelectronics industry. Since the lattice mismatch of CaF_2 on Si(111) is < 1%, it is a reasonable candidate as an epitaxial insulating film. Since STM relies on conduction between tip and sample, imaging an insulating system such as CaF_2, which has a bulk band gap of 12 eV, requires comment. It was found that the energetic positions of electronic states for sub-monolayers of the deposition product of CaF_2 and of CaF_2 multilayers have permitted STM imaging of the thin film in various stages of growth. In the sub-monolayer regime, different structures were observed as a function of coverage and deposition temperature, with row-like structures dominating, as shown in Fig. 100. Low-corrugation 1 × 1 structures that exhibit trapped negative charge begin to form just above 1 ML. Multilayer deposition at ~750 °C produced the image shown in Fig. 100. In the initial stages of growth, STS reveals new occupied (unoccupied) bands which develop at 1.3 eV below E_F (1.2 eV above E_F) corresponding to Ca–Si bonding (antibonding) levels, and this interface structure is a small-gap insulator which is easily imaged. The rows are C-derived features of dissociated CaF_2, most likely Ca^+ species of CaF bonded to Si adatoms. In the multilayer regime, E_F is pinned near the top of the Si valence band and the top of the CaF_2 valence band is 8.5 eV below E_F. It was therefore possible to tunnel into empty states of the insulating thin film at a bias of ~3.5 eV to obtain surface topographic information. Images of ~15 Å-thick CaF_2 layers reveal rough films.

The effect on an STM image of an atomic adsorbate on a semiconductor surface can be rather non-local due to the electrostatic effect of the local charge on band bending. This was first recognized for extremely small coverages of O on terraces of cleaved GaAs(110) surfaces. Locally, an O adatom, an electronegative adsorbate, appears as a protrusion in an occupied state image, while an unoccupied state scan pro-

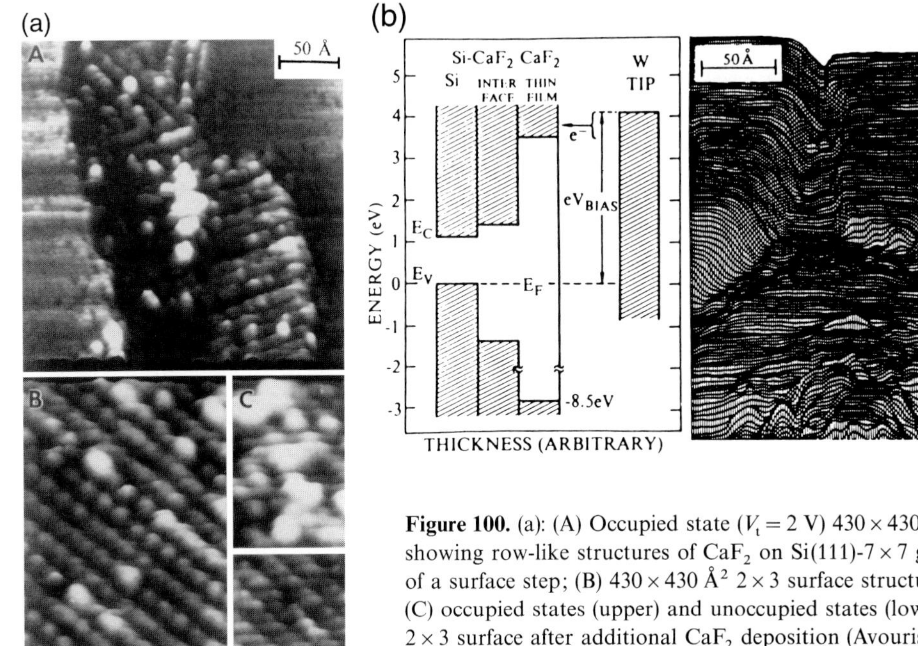

Figure 100. (a): (A) Occupied state ($V_t = 2$ V) 430×430 Å^2 image showing row-like structures of CaF_2 on Si(111)-7×7 growing out of a surface step; (B) 430×430 Å^2 2×3 surface structure ($V_t = 2$ V); (C) occupied states (upper) and unoccupied states (lower) of the 2×3 surface after additional CaF_2 deposition (Avouris and Wolkow, 1989 b). (b) Energy level diagram of the CaF_2/Si(111) interface with a topograph of a 15 Å thick CaF_2 film ($V_t = -3.5$ V).

duces a displaced local depression, as seen in Fig. 101. The topographic features completely heal only over a distance of tens of angstroms due to the electrostatic effect of the negative charge which fills the O 2p level that repels conduction band electrons and forms a depletion region around each adsorbate. This study demonstrated that local band bending caused by this coulomb field can extend for ≥ 50 A for Si-doped GaAs at 10^{18} cm^{-3}.

Metal Surfaces

Adsorption on metal surfaces is a site-specific phenomenon, and ordered adsorbate superstructures often form due to adsorbate–adsorbate interactions. As in the case of semiconductor surfaces, surface science techniques which average over large areas have provided much information on the structure of atomic and molecular adsorbates on metal surfaces. The use of STM of an ordered monolayer system to demonstrate the possibility of a direct structural view was first reported for the Rh(111)-$3 \times 3(C_6H_6 + 2CO)$ coadsorption system. Figure 102 a shows an empty-state image of individual benzene molecules in a 3×3 superlattice across a surface step. Due to the energetic position of the electronic states being imaged, CO molecules were not observed. The benzene molecules preferentially bond at sites near metal atoms forming the step edge. Notably, imaging with increased tunneling current at low biases produced corrugations that clearly showed the ring-like structure of benzene, as seen in Fig. 102 b. The electronic states that produce the observed corrugation were determined to be Rh-benzene states located just above E_F localized on the benzene π-lobes.

Thermally-activated processes which result in dissociative adsorption at room

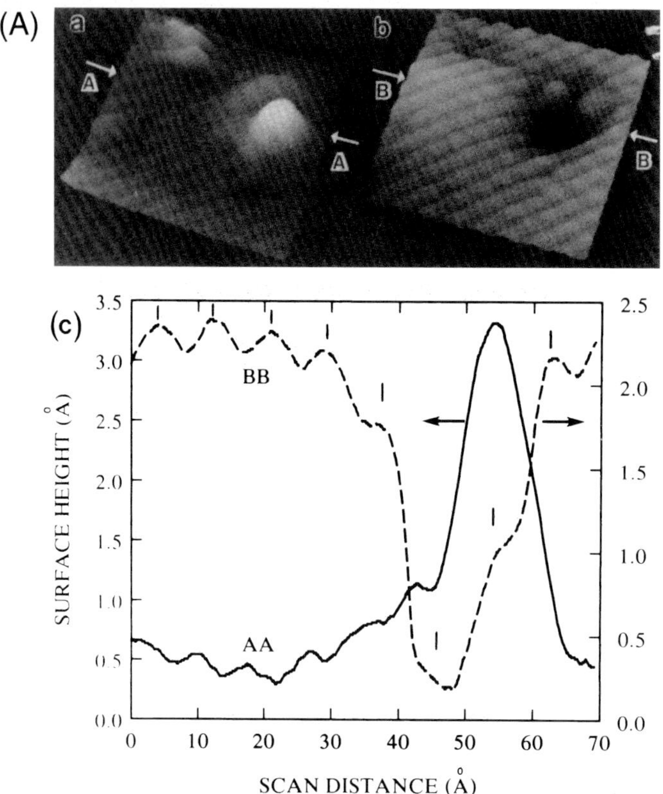

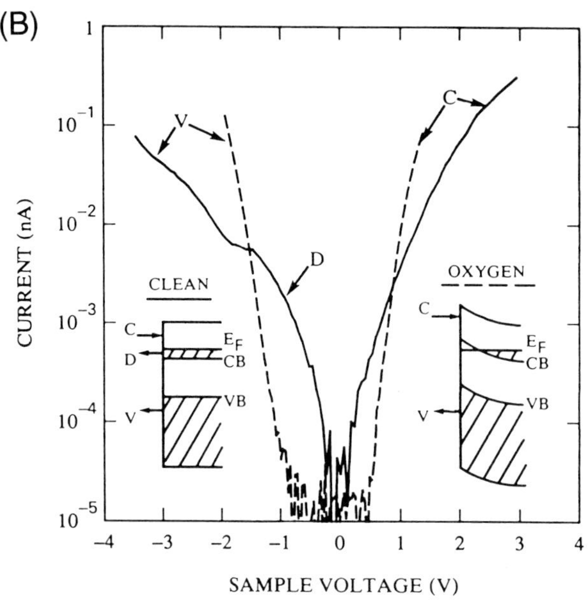

Figure 101. (A) STM images of an oxygen defect on GaAs(110) after 120×10^{-6} torr s exposure to O_2 acquired at (a) $V_t = +2.6$ V and (b) $V_t = -1.5$ V; (c) height contours through the oxygen defect and the corrugation of the Ga and As surface atoms. (B) $I-V$ measurements of the clean GaAs surface (solid line) and the oxygen-exposed surface (dashed line) indicating the effects of band bending in the vicinity of an adsorbate (Stroscio et al., 1987).

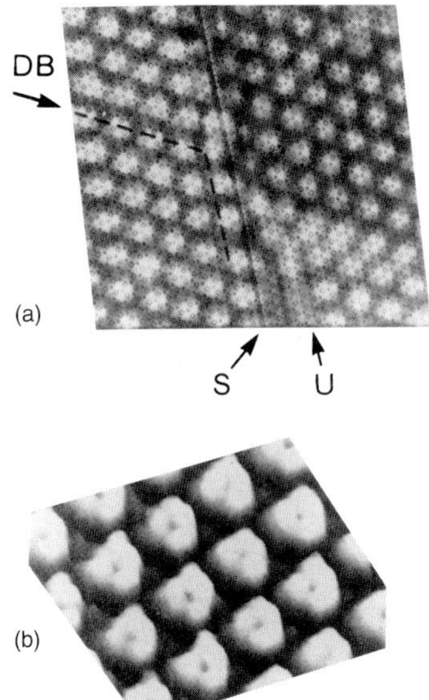

Figure 102. (a) Occupied states of the 3×3 C_6H_6 superlattice are imaged ($V_t = -1.25$ V and $I_t = 4$ nA) across an atomic step (S) with the Rh atoms indicated; a domain boundary (DB) is also visible. (b) Threefold structure and a depression in the center of the benzene ring are observed for $|V_t| < 0.5$ V (Ohtani et al., 1988).

temperature can be quenched at low temperatures. STM imaging commencing at low temperatures provides a means of identifying activation sites and long-range adsorbate–adsorbate interactions and variable-temperature imaging can reveal structural and electronic aspects of chemical reactivity. A recent study (Land et al., 1991) of ethylene (C_2H_4) on Pt(111) using a variable temperature STM operating between 150 K and 450 K demonstrates some of these points. From vibrational spectroscopy, it is known the C_2H_4 adsorption on Pt(111) at room temperature results in the irreversible formation of an ethylidyne species ($-C-CH_3$) bonded perpendicular to the surface. LEED has found that low temperature adsorption results in a (2×2) or in a multi-domain (2×1) C_2H_4 structure, and the $-C-CH_3$ also bonds in a (2×2) structure, as seen in Fig. 11-103. Imaging at 160 K showed a periodic array of ethylene molecules (~ 2 Å in size), although the periodicity was not completely correlated with the long-range order found by LEED. Warming to 350 K produces ethylidyne, but images obtained at room temperature show little or no structure. Only upon cooling back to ~ 180 K does the ethylidyne appear in the images, indicating that thermally-induced molecular motion must be taken into account. Since the structure is close-packed, the likely origin of this effect is low-frequency vibrational motion about fixed sites rather than molecular mobility over the surface. Fur-

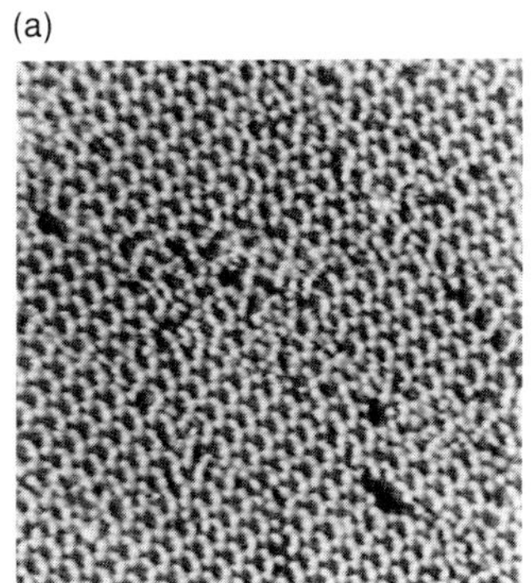

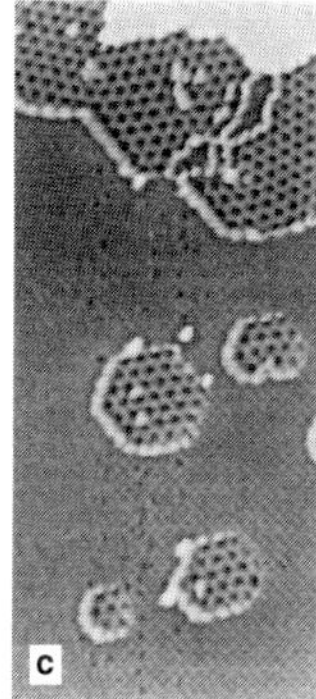

Figure 103. (a) 180×180 Å^2 topograph of C_2H_4 on Pt(111) obtained at 160 K after 2.5×10^{-6} torr s exposure. (b) STM topograph of ethylidyne-covered surface obtained at room temperature after annealing at 450 K. (c) Graphite islands observed in images recorded at room temperature after annealing at 1230 K (Land et al., 1991).

ther decomposition to carbon clusters or to a graphitic overlayer occurs upon heating to >700 K or to >1200 K, respectively. The graphitic overlayers exhibit a superstructure with a periodicity of ~ 22 Å, which suggests that the graphitic overlayer is higher-order commensurate with the substrate.

2.12 Electrochemistry at Liquid–Solid Interfaces

Besides the ability of ex situ characterization of surface topography before and after electrochemical processing (Gómez et al., 1986), STM is extremely well-suited as an in situ probe (or as an electrochemical electrode) for electrochemistry. For in situ studies, the tip and sample are immersed in solution along with the required electrode arrangements for electrochemical processing. In order to limit the tunneling current to the region, the tip shank is typically insulated with a glass coating such that only a few micrometers are exposed. We consider some aspects of the tunnel barrier and describe in situ measurements of electrochemical metal atom deposition and removal from the submonolayer to multilayer regimes. We note that the characterization of etching (Sonnenfeld et al., 1987) and the spectroscopy of electrolyte-induced states (Carlsson et al., 1990) at semiconductor surfaces have also been reported.

The barrier for tunneling in a liquid is different from that in vacuum or in air. In particular, measured apparent barrier heights, ϕ_a, are typically reduced, and achieving atomic resolution is possible but often more difficult. It has been proposed that chemical interactions between the tip and the liquid, or molecular packing of the liquid in the gap, can result in a change in the scaling of the tunneling gap. In addition, the more delocalized faradic current in the gap could account for reduced resolution. A recent theoretical discussion focused on the microscopic states of the liquid within the tunneling gap and tip-induced effects in order to describe the unique aspects of tunneling at the liquid–solid interface (Sass et al., 1991). The picture takes into account localized hydrated electronic states below the conduction band that can temporarily trap the tunneling electrons and tip-induced fluctations of the polar molecules that lower the tunneling barrier. In fact, a resonant tunneling process itself can account for the dramatic decrease in ϕ_a (Berthe and Halbritter, 1991).

After it was demonstrated that the STM could produce images in solution (Sonnenfeld and Hansma, 1986), real-time characterization of a surface during electrodeposition was reported for the Ag/graphite system (Sonnenfeld and Schardt, 1986). A three-electrode type cell was used with a graphite working electrode; a PtIr tip was used for imaging and not as an electrochemical electrode. Figure 104a shows the flat graphite surface before Ag plating. After two Ag plating cycles, during which time the tip was backed away from the sample, STM images of rolling hills reminiscent of vapor-deposited films were observed (Fig. 104b). Figure 104c shows that the Ag could be stripped from the surface by an oxidative voltage cycle to restore the clean graphite surface, which could be atomically-resolved; in this image, some Ag does remain as a circular island. Even after the equivalent of 300 ML Ag deposition, obtained from the volt-ammogram, islands next to the undisturbed substrate could be seen, while single Ag monolayers were never observed. This points to the fact that the pure island growth mode is operative for this system.

(a)

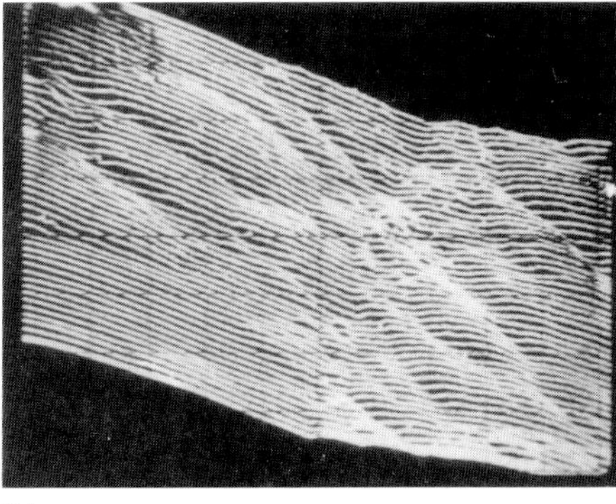

(b)

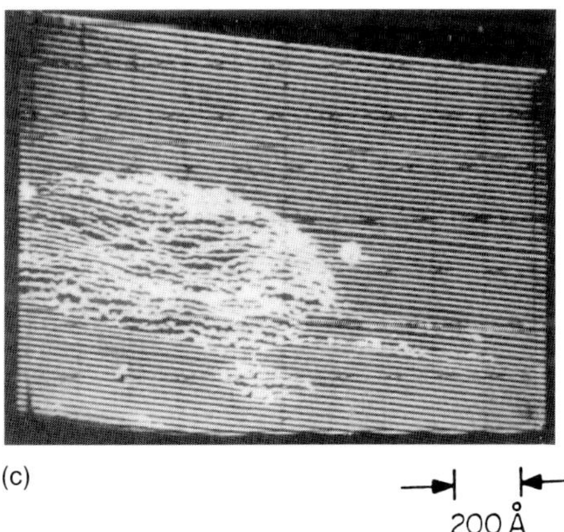

(c)

Figure 104. 900×1500 Å^2 STM images of graphite obtained with $V_{tip} = 100$ mV and $I_t = 10$ nA (a) before Ag plating, (b) after 59 mC electrodeposition of Ag, and (c) after removal of 59 mC Ag (Sonnenfeld and Schardt, 1986).

The initial stages of electro-crystallization of Au on Au(111) surfaces were studied (Schneir et al., 1988) using Au(111) surfaces prepared by growing epitaxial Au(111) films on mica, or by quenching gold wire in air after annealing in an oxy-acetylene torch flame. While atomic-resolution on this close-packed face was not obtained, images obtained in water prior to electroposition showed flat surfaces. In contrast, surfaces imaged in a $KAu(Cn)_2$-based electroplating solution exhibited steps and terraces apparently due to roughening (~ 10 Å) that occurred before the flow of ionic current. Using an electroplating solution, features with 10 Å roughness appeared on the surface. Electroplating 100 Å of Au (with the tip retracted) produced hillocks 30 Å high and 200 Å wide. Deposition of 400 Å Au produced an increased number of hillocks ~ 50 Å high. Finally, deposition of 1600 Å Au produced hills from 20–100 Å in height.

Monolayer growth of metal atoms on single crystal surfaces was demonstrated in the underpotential deposition (UDP) of Ag on Au(111) surfaces (Hachiya and Itaya, 1992). The STM used a glass-shrouded Pt tip in an aqueous sulfuric acid solution, and Ag and Pt wires were used as reference- and counter-electrodes, respectively. Atomically-resolved images were obtained, and the structures of the Ag adlayers were found to be particularly interesting. In comparison with the clean surface (Fig. 105 a), the first Ag layer produced, in an anodic scan from 0.5 V, was a $(\sqrt{3} \times \sqrt{3})R\,30°$ structure (Fig. 105 b) which had formed over a large area of the surface. Beyond the anodic peak potential at 0.525 V, the image lost atomic resolution until the Ag atoms were desorbed from the Au(111) surface at 0.6 V.

A real-time STM study of the surface evolution of an Ag–Au alloy in perchloric

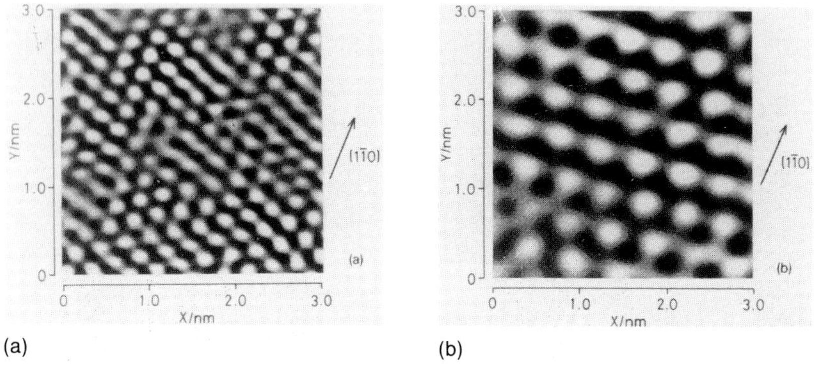

Figure 105. 30×30 Å^2 images of the Au(111) surface obtained with (a) an electrode potential of 0.5 V, showing the Au(111)1×1 structure and (b) an electrode potential of 0.5 V, showing a $\sqrt{3} \times \sqrt{3}$ Ag layer deposited on the surface (Hachiya and Itaya, 1992).

acid (0.1 M $HClO_4$) was undertaken to address the microscopic mechanisms of selective dissolution (de-alloying) and surface diffusion in the corrosion process (Oppenheim et al., 1991). Ag-deficient alloys were prepared by vapor co-deposition of Ag and Au onto heated mica substrates. A sequence of STM images of $Ag_{0.07}Au_{0.93}$ obtained using PtIr tips, as shown in Fig. 106, shows that the starting surface (a) roughens (b) and forms monolayer-deep pits and vacancy clusters (c). Near the ledges on the surface, pits interact with the ledges on the surface, giving a noticeable curvature to the surface (d). More dilute alloys exhibit less roughening, while higher concentrations of Ag give larger amounts of roughening. In addition, the potential for Ag dissolution is higher in the alloys than for pure Ag, similar to what is observed for removing an underpotential-deposited Ag monolayer from Au(111) (Hachiya and

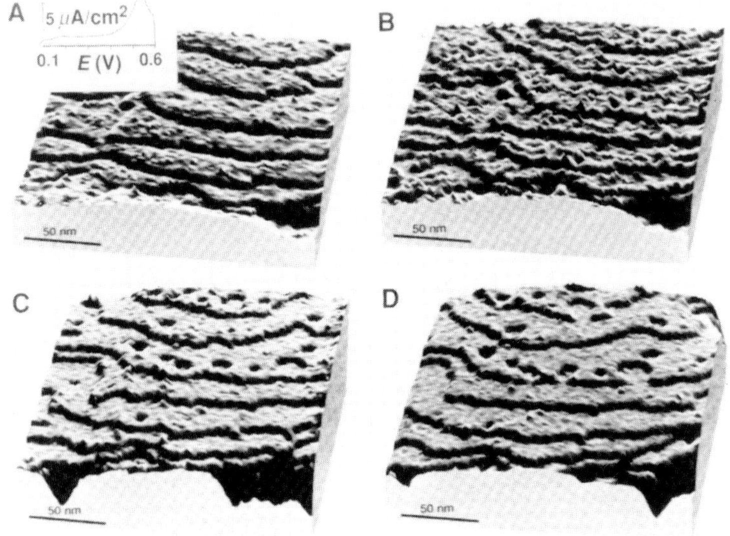

Figure 106. Time sequence of STM images of the $Ag_{0.07}Au_{0.93}$ sample in 0.1 M $HClO_4$ obtained (A) at the onset of the application of an electrochemical potential, and after scanning the electrochemical potential to 0.60 V, (B) after 10 min, (C) after 51 min, and (D) after 86 min (Oppenheim et al., 1991).

Itaya, 1992). The selective dissolution process was explained in the kink-step-terrace model. As the number of Ag atoms at kinks and steps is reduced, it is necessary to apply an electrochemical overpotential to remove terrace Ag atoms, creating monatomic pits. If the Ag concentration is below the 3D percolation threshold, Ag cannot be removed from the bulk and the surface rearranges by vacancy diffusion.

2.13 Biological Systems

STM is clearly capable of obtaining atomic-scale images and spectroscopic information on organic molecules on surfaces. The fact that STM can characterize such materials in vacuum, in air, or in solution implies that entirely new classes of materials and problems can be addressed. One important class of problems includes the use of STM for biological molecules. One of the first STM studies of a biological system reported images of DNA (deoxyribonucleic acid) components (Baró et al., 1985), and this study attracted a great deal of interest. The prospect of in vivo characterization with a relatively simple instrument, as well as the possible nanoscale manipulation of biological materials, could be envisioned. In much of the work that has followed, physicists, chemists, and biologists have cooperated to formulate the basis for studying biological systems with STM by focusing on sample preparation, characterization, and image analysis. We note that, since many of these systems are non-conducting and typically macromolecular features which are of primary interest, there has been a shift to the use of more recently developed AFM for these systems, as discussed in Sec. 3.1.6. However, the imaging and spectroscopic capabilities of STM are suitable for many biological systems.

Of the questions and complications which arise, the most fundamental are relevant for electron microscopy as well. First, how well does the sample represent the actual biological system? Second, what effect does the probe or preparation method have on the system being studied? Studies of the topography of a variety of biomolecules, including DNA (Baró et al., 1985; Lindsay and Barris, 1988; Beebe Jr. et al., 1989; Lindsay et al., 1989; Bendixen et al., 1990; Miles et al., 1990; Salmeron et al., 1990; Selci et al., 1990; Thundat et al., 1990; Cricenti et al., 1991; Youngquist et al., 1991), proteins (Guckenberger et al., 1989; Hameroff et al., 1990; Jericho et al., 1990; Miles et al., 1990; Hörber et al., 1991), enzymes (Elings et al., 1990; Masai et al., 1992), membranes (Guckenberger et al., 1991), and chloroplasts (Mainsbridge and Thundat, 1991; Dahn et al., 1992), as well as biomolecule – biomaterial interfaces (Emch et al., 1990), have been reported. These pioneering studies have not only shed light on particular biological systems but have also helped define and solve the experimental difficulties associated with STM analyses of these systems.

Some of the limitations associated with imaging biomolecules on surfaces with STM have been discussed in the literature (Salmeron et al., 1990). The most serious problem is the degree of electron conduction through an insulating biomolecule. A straightforward but somewhat unappealing solution to this problem is to prepare metal-coated samples, as is commonly done in electron microscopy. However, even ultrathin coatings reduce the ultimate spatial resolution that can be achieved. Furthermore, environments similar to those of the living system are desired, but such samples may be modified irreversibly. Thin molecular layers on conducting substrates, such as graphite and metals, have in fact, produced high-quality (atomic-scale) imaging of non-conductive materials. Tunneling in this regime has been modeled in one-dimension (Salmeron et al., 1990) as conduction through two barriers with the biomolecule residing in a central well. The first tunnel barrier is tip-to-molecule and the second represents molecule-to-surface. Calculating the tunneling probability standard methods, the presence of the molecule changes the phase of the electron wavefunction and provides conditions for resonant tunneling.

Another problem involves the strength of the interaction of the biomolecule with the substrate. This interaction is usually weak. Then, if the attractive tip–molecule interaction is greater than the tip–substrate interaction, the tip can displace the molecule along the surface. Stronger attachment points appear to occur at steps or defects on the substrate surface, as observed by a higher density of material in these regions. Chemical fixation, e.g. by adsorbing the biomolecules on a pre-deposited or co-deposited attractive surface layer, such as a lipid, is another technique to limit surface motion (Masai et al., 1992). One technique introduced to reduce tip-induced motion was to use STM in a "hopping" mode, in which the tip is scanned away from the surface for a majority of the time and only images during a part of the scanning sequence. We note that it has been suggested (Salmeron et al., 1990) that the advantage of a metal coating is that it fixes the molecule rather than that it provides conductivity.

We discuss a few of the highlights of the work on DNA; studies of chloroplasts, where fine structure on micrometer-sized objects was observed; and images of protein complexes anchored on lipid bilayers. STM imaging of the double-helix structure of DNA has been performed due to the broad scientific interest in the helical structures of the order of tens of angstroms. Such measurements have been performed on air-dried samples as well as on samples in solution.

A DNA strand lying flat on a substrate has dimensions suitable for imaging by STM – ~ 20 Å diameter and ~ 400 Å in length. The first reported attempt at imaging air-dried bacteriophage *ϕ29* DNA particles deposited on a graphite substrate (Baró et al., 1985) produced structural features comparable in size to analogous samples imaged by electron microscopy. Although vertical resolutions of ~ 1 Å were obtained, it was recognized at this point that the mechanism for tunneling a propagation of electrons through thick sulating samples is markedly different vacuum tunneling on conducting su where images of individual atoms a tained.

More recent STM images (Beebe Jr. et al., 1989), shown in Fig. 107, give remarkably clear structures for air-dried calf thymus DNA. Image (a) and the processed image (b) are depicted in (c) as a double-stranded DNA particle which is curled over itself at the left. These high resolution images exhibit corrugations along the axis containing the DNA duplex. Periodicities of the corrugations along a particle (~ 49 Å) correspond to the helix pitch. The image in (b) almost isolates major and minor grooves of the helix. The observed height (20–30 Å) agrees with the expected value, while the observed width (60 Å) was greater than expected (20–30 Å) due to the convolution of the tip structure with the molecular structure. The large variation in the measured DNA pitch (up to ~ 37%) was attributed to dehydration of the sample during preparation. STM images of calf thymus DNA particles on a gold surface and submerged in water were found to re-

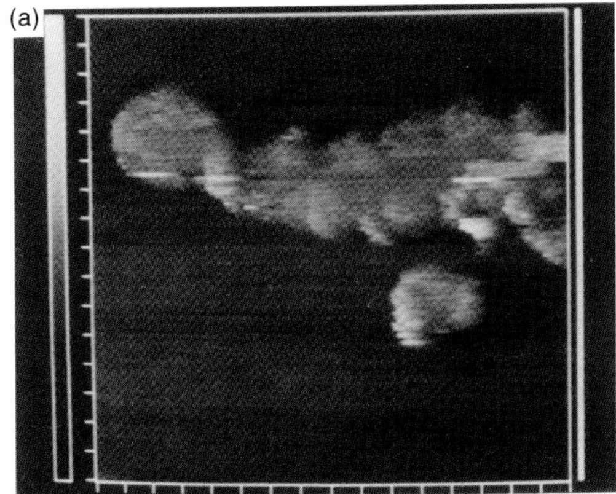

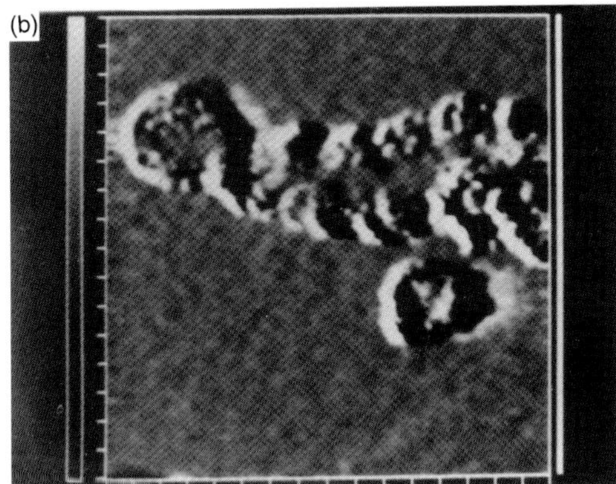

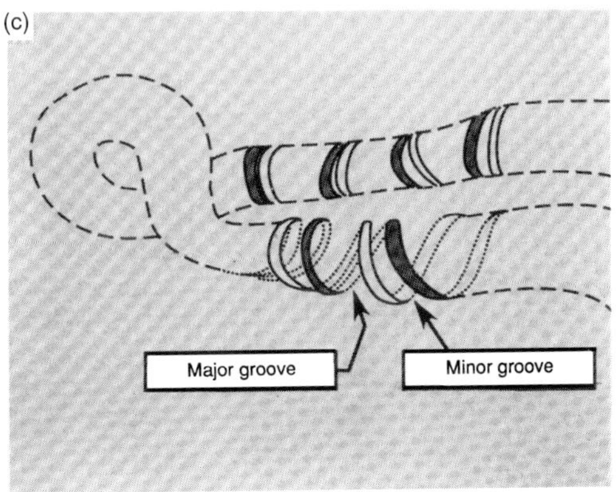

Figure 107. (a) 400×400 Å^2 image of air-dried DNA on a graphite substrate obtained with $V_t = \approx +100$ mV and $I_t \approx 3.3$ nA. (b) Processed STM image. (c) Diagram of the DNA structure (Beebe Jr. et al., 1989).

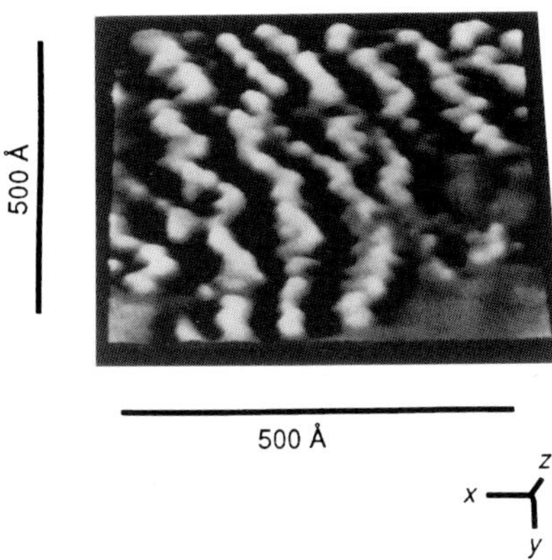

Figure 108. 620×620 Å^2 STM image of DNA on a gold surface obtained in distilled water. The three complete fragments on the left side of the image exhibit distortions not seen in images of air-dried samples (Lindsay et al., 1989).

sult in a more regular pitch (Lindsay et al., 1989). However, DNA strands in solution, shown in Fig. 108, were found to exhibit a large amount of distortion (bending and curving) unless the packing density was increased; this occurs due to tip-induced molecular deformation during the scan.

A recent study (Dahn et al., 1992) comparing variously prepared chloroplast samples addresses several sample preparation issues. The distinction between "unbroken" (whole but with membrane disruption) and "intact" (no membrane disruption) chloroplasts was noted when a comparison was made between images of gold-coated chloroplasts examined in air and bare specimens characterized in air and in solution. The coated sample, shown in Fig. 109, shows a whole chloroplast a few micrometers in lateral dimension and < 1 µm high. Grooves of the order of 0.5 µm were related to connections between the chloroplast envelope and the inner membranes. Finer structures on the scale of ≤ 10 nm were considered to be artifacts of the metal film. It was found that imaging uncoated samples in air was im-

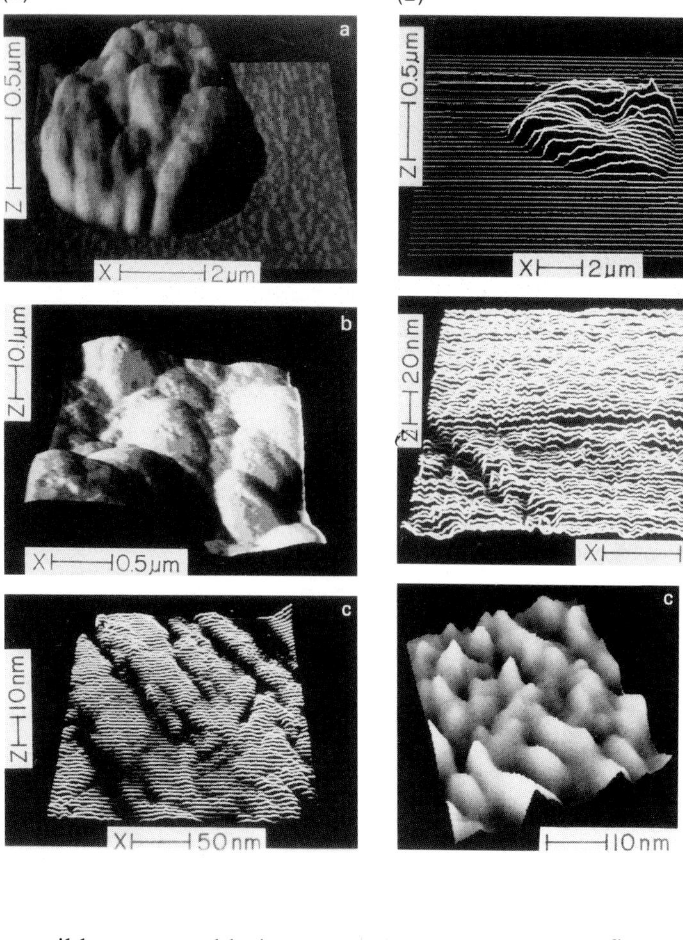

Figure 109. (A) STM images of gold-coated chloroplast in air. (a) Entire chloroplast; (b) a magnified image of the structure on top of the chloroplast; (c) higher magnification of the image showing the structure due to the gold coating. (B) STM images of uncoated chloroplast in distilled water. (a) Entire chloroplast; (b) a magnified image of the structure on top showing a 10 nm deep groove which has been identified as a membrane contact site; (c) higher resolution image showing surface corrugations < 10 nm in lateral dimension (Dahn et al., 1992).

possible, presumably because dehydration renders the samples to be extremely fragile and/or reduces the conductivity to an unacceptably low level. Imaging samples in solution (distilled water) required that the current be held at ≤ 0.1 nA, so that the tip would be close enough for resolution without damage to the samples. An image of an intact chloroplast is shown in Fig. 109 B, a. Chloroplasts were observed to be flatter (≤ 0.3 µm high) than the coated samples. Grooves observed on the chloroplasts in solution were analogous to those seen on the coated specimens. The significance of STM imaging involves the fine detail on the < 10 nm scale, which demonstrated the possibility of observing very fine scale structure on uncoated biological specimens. It has been proposed that the 10 nm features correspond to protein chains at the surface (Mainsbridge and Thundat, 1991).

ATP synthase (F_0F_1-ATPase) is an enzyme involved in proton transport in energy-transducing membranes. An STM study of reconstructed sub-unit structures of F_0F_1-ATPase, anchored in a lipid bilayer on a HOPG substrate (Masai et al., 1992) demonstrates the ability of STM to image fine surface details of such sub-unit complexes and illustrates the connection between TEM (transmission electron microscope) and STM images. TEM images show that two-dimensional crystals of hexagonal structures of the TF_1 sub-unit form as shown in Fig. 110b. STM images of these biomolecules, shown in Fig. 110c, d, exhibit approximately hexagonal structures of ~10 nm lateral dimension each with a corrugation in the center. This is consistent with the picture of the F_0F_1-ATPase layer in the lipid bilayer (Fig. 110a).

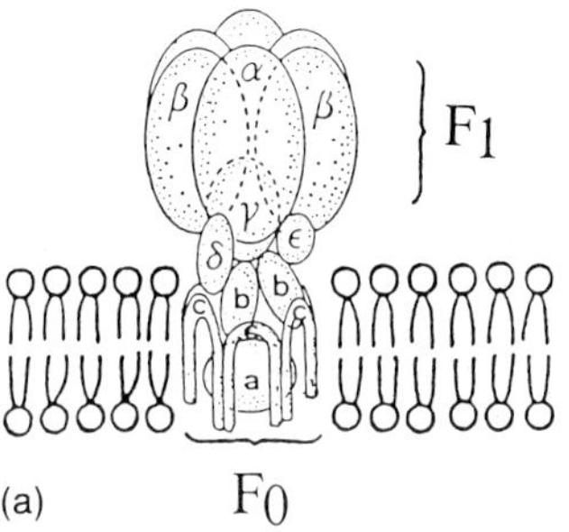

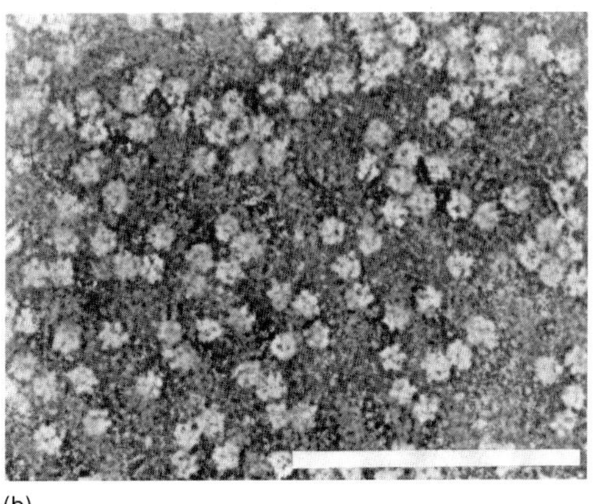

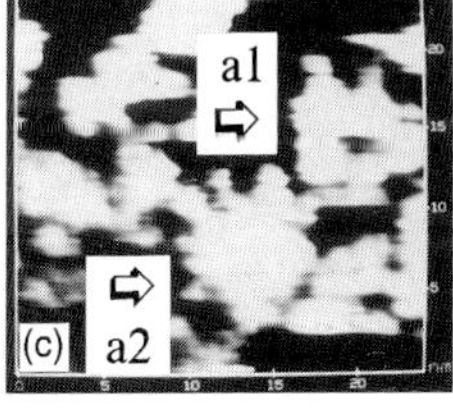

Figure 110. (a) Drawing of F_0F_1-ATPase embedded in a lipid bilayer. (b) A TEM image of TF_1 indicating hexagonal structures with a depression at their centers. (c) STM image of TF_1 on graphite showing structures similar to those observed by TEM. (d) Processed hexagonal structure showing center depression (Masai et al., 1992).

2.14 Metrological Applications

The use of the STM as a metrological tool of very high (three-dimensional) spatial resolution is rooted in the original concept and goals of the original Topografiner (Young et al., 1972) described in an earlier section. While the achievement of atomic or near-atomic resolution, well-characterized surfaces does not necessarily mean that the STM can be used on any surface, the possibility of obtaining even nanometer resolution on a surface provides an important complement to other microscopies. Present-day STMs can image large regions separated by very narrow channels. Instruments capable of scanning up to tens of micrometers allow direct connections to commonly-used forms of microscopies. In addition, the STM provides direct digital topographs that can be used for quantitative analysis. This provides the ability, for example, to determine statistical variations in surface heights as a direct measure of surface roughness. In some cases, STMs are mated to scanning electron microscopes (SEMs) (Gerber et al., 1986), or are part of a comprehensive measurement system, such as the molecular measuring machine (M^3) developed at the National Institute of Standards and Technology (Teague, 1989). The latter was designed to span length scales of tens of millimeters with sub-nanometer point-to-point precision.

Measuring surface roughness and the distribution of island sizes on a surface (Denley, 1990; Habib et al., 1990; Reiss et al., 1990) requires that the tip contribution to the acquired image be removed by deconvolution. Depending on the tip size and shape, a limited number of regions not accessible to the tip may remain after deconvolution. Surface roughness can be quantified by defining the function $r(h)\,\mathrm{d}h$, which is the ratio of those areas of $i(x, y)$ at heights between h and $h + \mathrm{d}h$ with respect to the entire surface, where $i(x, y)$ defines the STM image (Reiss et al., 1990). The mean surface height, h_0, for a flat surface is then written as

$$h_0 = \int_{-\infty}^{+\infty} h\, r(h)\, \mathrm{d}h \tag{20}$$

Normalized to h_0, $r(h)$ represents the distribution of roughness above and below the mean. Images of rough Au and Ni films are shown in Fig. 111 a, b with a distribution of $r(h)$ quantitatively indicating the greater roughness of the Au film (Fig. 111 c). The distribution of lateral structures can be obtained by the autocorrelation function $\mathrm{acf}(\boldsymbol{a})$ over the surface S, written as

$$\mathrm{acf}(\boldsymbol{a}) = S^{-1} \int_S i(\boldsymbol{r})\, i(\boldsymbol{r}+\boldsymbol{a})\, \mathrm{d}^2\boldsymbol{r} \tag{21}$$

The autocorrelation function for $a = 0$ gives the square of the root mean square (rms) roughness. If the distribution of island sizes follows a white noise law, there is an exponential decrease in $\mathrm{acf}(\boldsymbol{a})$ for increasing values of a. The mean island diameter is extracted by noting distances between peaks in $\mathrm{acf}(\boldsymbol{a})$, as indicated in Fig. 111 d.

Microfabricated structures are typically imaged with SEM. The addition of STM, operating in a large-field mode, is very appealing as it offers the capability of obtaining a significant improvement in detail in three dimensions. In some cases, SEM and STM have been used together with the SEM used to position the tip in the region of interest, as for the DC SQUID microbridge (Anders et al., 1988) illustrated in Fig. 112.

Other microfabricated patterns have also been studied to determine topography

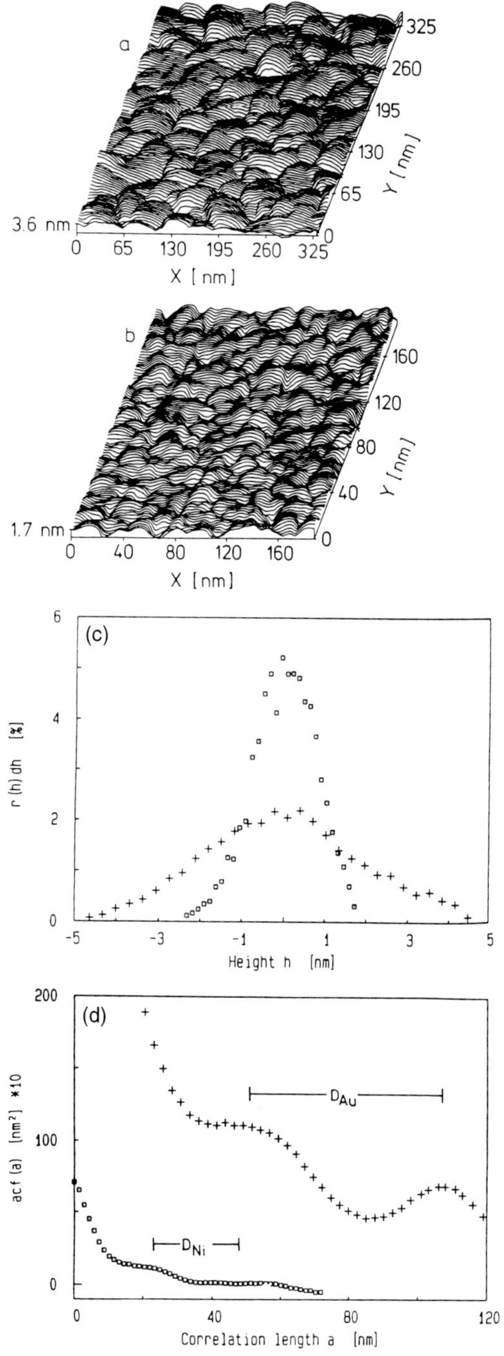

Figure 111. STM topographs of (a) a 20 nm thick Au film and (b) a 20 nm thick Ni film deposited on polished glass at room temperature; (c) distributions of roughness, $r(h)\,\mathrm{d}h$, for the Au(+) and Ni(□) films; and (d) autocorrelation functions, $\mathrm{acf}(\boldsymbol{a})$, for the Au(+) and Ni(□) films (Reiss et al., 1990).

and channel structures. Figure 113 compares SEM and STM images of an Au-coated grid pattern obtained by ion-enhanced chemical etching (Okayama et al., 1988). A 50 keV focused and rastered Ga^+ liquid metal ion source (LMIS) ion beam was used to bombard an Si surface, and chemical etching followed.

The microtopography and nanotopography of a surface is crucial in many applications, such as for high-precision optical components and disk drive surfaces. STM is ideal to characterize the surface roughness of machined or ground surfaces in areas where such a finish is crucial. This type of information can guide the development of tool and machining specifications for such components. Figure 114 shows an STM image of an individual turn mark on a diamond-turned Al substrate (coated with an AuPd film) to be used for subsequent magnetic film deposition for a high-capacity hard disc drive (Gehrtz et al., 1988). The main advantage of an STM image over the corresponding SEM image is in the precise characterization of the height of the turn mark, which in this case is of the order of 20 nm.

(a)

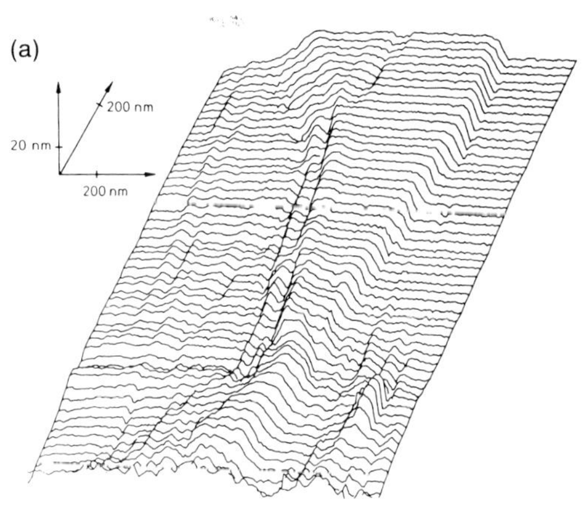

(b)

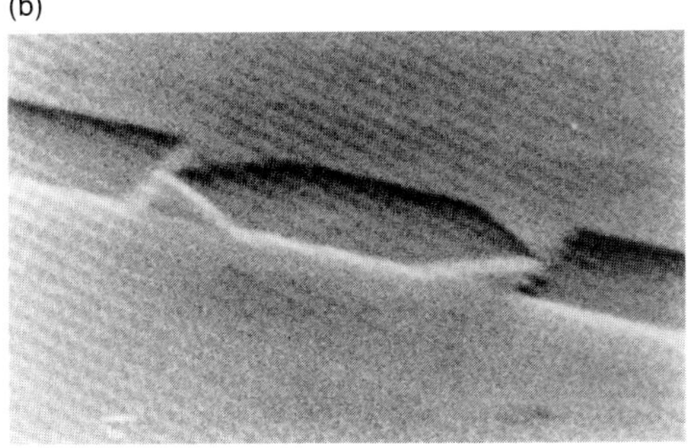

Figure 112. (a) STM image of a DC SQUID microbridge with corresponding SEM image (b) (Anders et al., 1988).

(a)

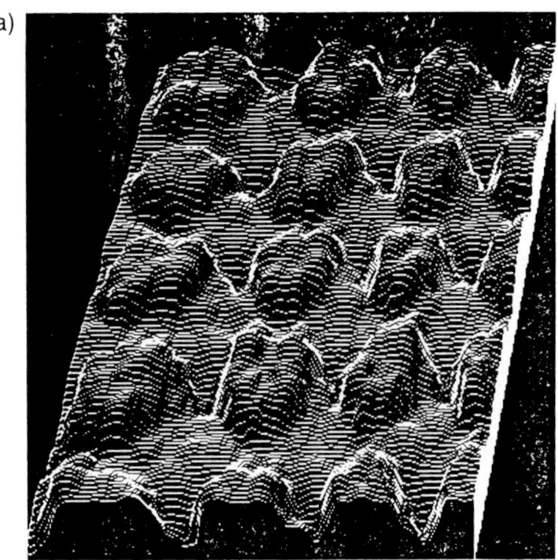

(b)

Figure 113. (a) STM image of a sub-micron grid pattern formed by Ga^+ ion-beam-enhanced chemical etching of an Si surface and (b) corresponding SEM image (Okayama et al., 1988).

(b)

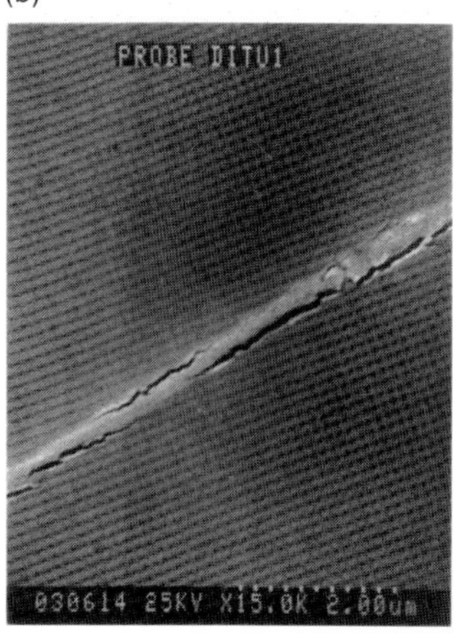

Figure 114. (a) STM image of a turn mark on a diamond-turned Al substrate and (b) corresponding SEM image (Gehrtz et al., 1988).

3 Atomic Force Microscopy

The previous discussion of the scanning tunneling microscope (STM) demonstrated that tip interactions with a surface can lead to measurable deformations, which can result in anomalous images or, at least, must be taken into consideration in a structural analysis (Tersoff, 1986). Such tip–surface interactions can even be used to inscribe the surface. Besides repulsive *contact forces*, both attractive and repulsive *non-contact forces* (van der Waals, electrostatic, and magnetostatic interactions) can influence the tip far beyond typical STM tip–sample separations. The atomic force microscope (AFM), invented by Binnig, Quate, and Gerber (Binnig et al., 1986) in 1986, was developed to exploit contact and non-contact forces for imaging surface topology and to study new physical phenomena at microscopic dimensions. The AFM is a major extension of STM and has borrowed STM technology, including the use of piezoelectric transducers for sub-angstrom motion and implementation of various feedback schemes. In the AFM, the force-transducer is a deflecting cantilever on which a sharp tip is mounted.

As a topographic imaging technique, the AFM may be viewed as a stylus profilimeter; atomic resolution is obtained by reducing the contact force of a commercial profilimeter ($\sim 10^{-4}$ N) to below 10^{-9} N, which is less than most interatomic forces, limiting tip-induced surface deformation and, consequently, minimizing the contact area to allow imaging of single atom. Prior to the development of the AFM, other short-ranged force measurement techniques had been developed such as the "surface force apparatus" (Israelachvili and Adams, 1978). Forces between two surfaces which confine nanometer-thick organic films can be measured with a vertical accuracy of ~ 0.2 nm and a force sensitivity of $\sim 10^{-7}$ N. With the AFM, enhanced force sensitivity and the introduction of a scanning probe makes it possible to image on an atomic scale and to develop a nanoscale basis for long-ranged surface forces and other phenomena, such as adhesion and tribology.

The heart of the AFM is the cantilever and tip assembly, which is scanned with respect to the surface, as shown schematically in Fig. 115. For *contact imaging* the sample is scanned beneath the atomically-sharp tip mounted on the cantilever providing a repulsive force. Atomic resolu-

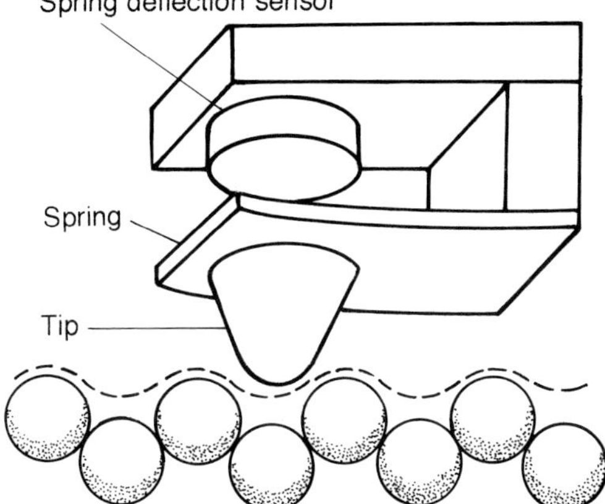

Figure 115. Schematic representation of the atomic force microscopy technique for imaging surface atomic corrugation. A tip of atomic dimensions is mounted on a cantilever and is in contact with the surface. In the variable force mode, the cantilever moves over the surface, and the spring deflection due to the corrugation is plotted as a function of lateral position (Rugar and Hansma, 1990).

tion can be obtained by very light contact and measuring the deflection of the cantilever due to the repulsion of contacting atomic shells of the tip and the sample. The image represents a picture of the total charge density, whereby the corrugation can be modeled in terms of hard spheres in contact. Considering the macroscopic size of most tips it is believed that atom-sized asperities that contact the surface produce the high resolution images. It is most appropriate to maintain a small constant force on the sample for the entire image to insure small uniform deformation. This can be accomplished by using feedback to demand constant cantilever deflection by compensating the sample z-position. A plot of $z(x, y)$ then represents a constant force topographic image. The magnitude of the imaging force is set by adjusting the dc-bias on the sample z-piezo. Most AFMs operate using this arrangement although variations in the details are common.

Contact forces in imaging should be small enough so that the sample surface is not appreciably disturbed. The bonding of atoms at surfaces may involve either strong ionic forces, medium-strength covalent bonding forces, or weak van der Waals forces. The magnitude of these forces can be estimated (Binnig et al., 1986) by considering a typical potential energy function of an interatomic bond. If we estimate the ionic bonding energy $U \leq 10$ eV and a van der Waals bonding energy of $U \leq 10$ meV and take the repulsive contact force as acting through a distance $\Delta x \approx 0.2$ Å, the interatomic force, $F = -\Delta U/\Delta x$, would be $\leq 10^{-7}$ N for ionic bonds and $\leq 10^{-11}$ N for van der Waals bond. These forces are also consistent with typical vibrational frequencies, $\omega = (k_{\text{bond}}/m_{\text{a}})^{1/2}$, where k_{bond} is the interatomic force constant and m_{a} is the atomic mass.

These estimates define the requirements for the force imparted by the probe, i.e., the force constant of the cantilever. With the force constant < 0.1 N/m, even a 10 nm loading deflection creates a force of $< 10^{-9}$ N. Such cantilevers can readily be constructed. For example, the first AFM (Binnig et al., 1986) used a diamond chip attached to a 0.8 mm × 0.25 mm × 25 μm gold foil; the spring constant was ~ 0.01 N/m with a resonance frequency $f_{\text{r}} = 2\pi\,(k/m)^{1/2} \approx 2$ kHz. Besides foils, metal wires and carbon-fiber bridges have also been used successfully. A softer spring would impart less force for a given deflection, and as such would reduce the forces on the sample. To maintain a high resonance frequency and good vibration isolation, however, a reduction in the force constant must be coupled with a reduction in mass to preserve the ratio k/m. Silicon device technology has introduced microfabrication techniques to produce silicon, silicon oxide, or silicon nitride microcantilevers of extremely small dimensions, on the scale of 100 μm × 100 μm × 1 μm thick (Akamine et al., 1990). Microcantilevers exhibit force constants around 0.1 N/m and resonance frequencies as high as 100 kHz. Cantilevers of the type shown in Fig. 116 achieve the desired design goal – a large k and a small cantilever mass gives a high resonance frequency (up to 100 kHz) for superior external vibration rejection.

Considering the magnitudes of forces being discussed, measuring a very small cantilever deflection is crucial for a high-sensitivity force measurement. The first AFM employed an integral STM tip on the backside of the gold cantilever to measure its deflection (Binnig et al., 1986). Over a short period of time, optical levers (Barrett and Quate, 1990), optical interferometry (Erlandsson et al., 1988), and capacitance probes (Göddenhenrich et al., 1990a) have emerged to detect tip deflection along the z axis or even along two orthogonal axes (Neubauer et al., 1990). Figure 117 presents diagrams of two AFMs. Figure 117a illustrates how the deflection of the cantilever utilizes a crossed fiber which holds a tungsten tip (a). Deflection is measured with great precision by an STM tunneling tip sensing the back surface of the AFM tip. Zero deflection is set when the sample is retracted far from the tip; at that point, the STM tip is advanced to tunnel from behind the cantilever, and a force measurement can commence. Figure 117b shows an arrangement where the deflection is measured with an optical lever (b). Here, laser light is reflected off the back of the cantilever and is sensed by a position-sensitive detector.

In *contact imaging*, the measurement of cantilever deflection is performed directly (quasi-statically) as the tip is being scanned over the surface. The AFM can also image and measure long-ranged interactions with the tip – van der Waals, electrostatic, magnetic – at relatively large distances from the

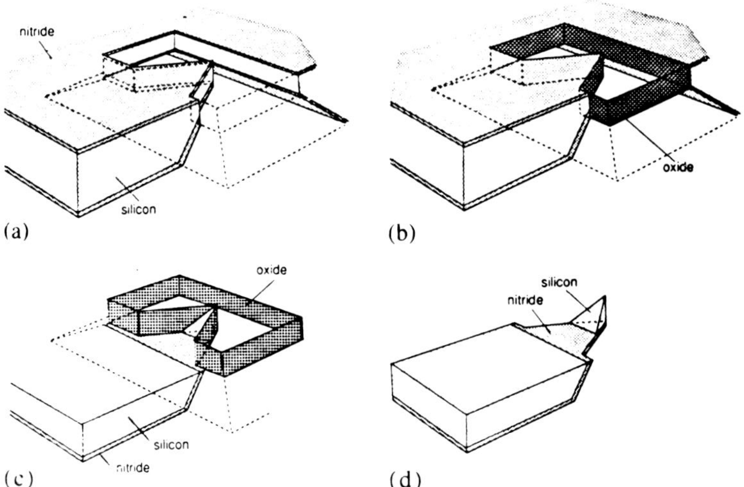

Figure 116. Process for fabricating a cantilever with an integral tip using an Si wafer with Si_3N_4 films on both sides: (a) the cantilever is produced using photolithographic techniques; (b) the exposed Si surfaces are oxidized; (c) the nitride layer is removed from the front of the wafer and the exposed Si is etched; (d) the oxide is selectively removed leaving a cantilever with an Si tip exposed. For a typical cantilever with a thickness of 0.7 μm, the force constant is ~0.6 N/m and the resonance frequency is ~80 kHz (Akamine et al., 1990).

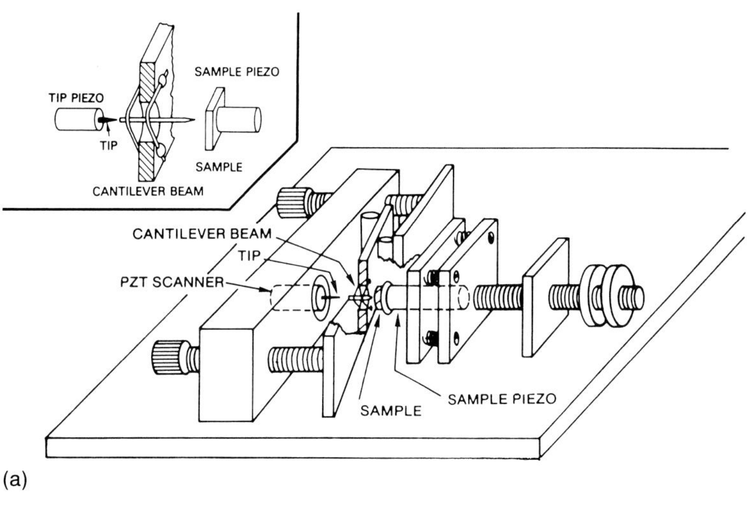

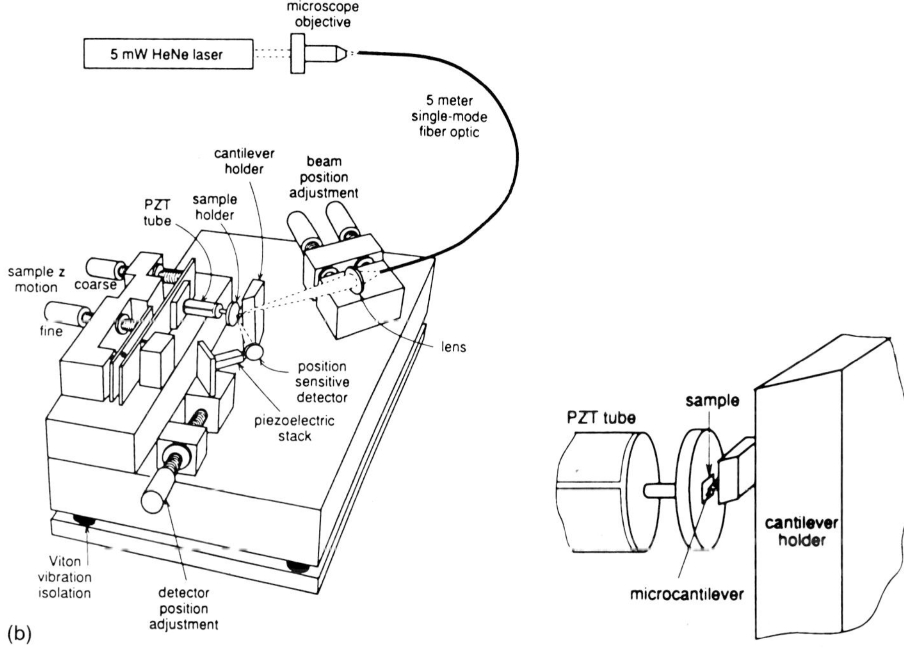

Figure 117. (a) Diagram of a double-crossed-fiber cantilever beam AFM. Here an etched tungsten tip is attached to the cantilever for microindentation measurements; such an arrangement gives a cantilever force constant of ~ 50 N/m. The STM tip tunnels to the back of the W tip to monitor its motion (Burnham and Colton, 1989). (b) Diagram of an optical lever AFM which operates by measuring the movement of a light beam by a position-sensitive detector after the light is reflected from a mirror on the back of the cantilever. With a cantilever motion sensitivity of 0.1 nm the force sensitivity is $\sim 10^{-9}$ N with a force constant of 10 N/m (Barrett and Quate, 1990).

sample by *non-contact imaging*. The lateral resolution is substantially decreased in the latter imaging mode, but, for the applications of interest, the AFM is far superior to alternative methods. Figure 118 shows plots of force as a function of distance using representative values for van der Waals, electrostatic, and magnetic forces. These forces are rather small at typical probe–sample separations of 10–100 nm; forces $< 10^{-10}$ N are a good deal smaller than the repulsive contact force for atomic resolution imaging. To measure these forces requires instrumental sensitivities as high as $\sim 10^{-13}$ N and beyond; such performance can only be obtained with resonance enhancement techniques.

Several schemes based on driving the cantilever at (or near) its mechanical resonance frequency and detecting force-gradient-induced deviations from the resonance condition have been developed (Sarid and Elings, 1991). If a cantilever is driven at its free resonance frequency by piezoelectric elements or thermally (Umeda et al., 1991), immersion in a force field results in a deviation of the resonance frequency, which can be detected by measuring changes in parameters of the mechanical system – a resonance frequency shift, a change in amplitude of the cantilever driven at fixed frequency, or a phase change. The instantaneous position of the cantilever can be measured by tunneling, capacitance, or optical detection methods (homodyne, heterodyne, laser-diode feedback, polarization, or deflection). The transfer function of the system is directly related to the force gradient.

To illustrate the use of resonance techniques, we consider a cantilever with force constant, k, oscillating at its resonance frequency, $\omega_0 = (k/m_{\mathrm{eff}})^{1/2}$, where the effective mass, m_{eff}, depends on the mass distribution and geometry of the tip/cantilever assembly (Martin et al., 1987). Assume that the cantilever is driven at ω_0 by a piezoelectric transducer. If the tip is placed in a force field, $\boldsymbol{f}$, then

$$k \Rightarrow k' = k - \frac{\partial f}{\partial z} \tag{22}$$

where $\partial f/\partial z$ is the force gradient and z is the vertical separation of the tip and the sample. The negative sign signifies an attractive force. The effective force constant is softer than for the free cantilever, thereby reducing the resonance frequency to ω_0',

$$\omega_0 \Rightarrow \omega_0' = \frac{k'}{m_{\mathrm{eff}}} < \omega_0 \tag{23}$$

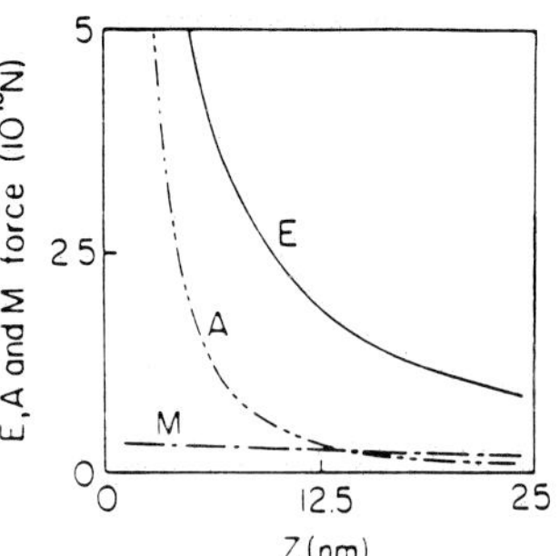

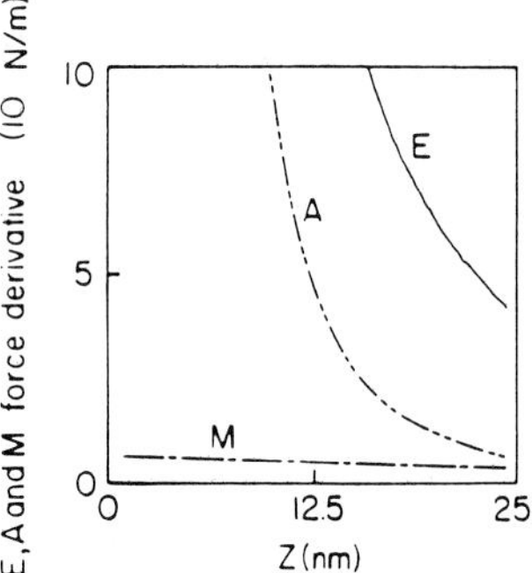

Figure 118. $F(z)$ and $\partial F(z)/\partial z$ extend for a significant distance from the surface, as illustrated for electric (E), van der Waals (A), and magnetic (M) forces. E and A are estimated for a sphere–plane interaction and M is estimated for a sphere–sphere interaction (Sarid and Elings, 1991).

The equation of motion for the system,

$$m\frac{\partial^2 z}{\partial t^2}+\gamma\frac{\partial z}{\partial t}+k(z-u)=f(z-g) \quad (24)$$

includes a damping term, $\gamma = m\,\omega_0/Q$, where Q is the quality factor, u is the position of the undeflected cantilever, g is the sample position, and z is the position of the deflected cantilever. Altough bent wires and foils can be used as cantilevers, microfabricated cantilevers give an increased quality factor which can be increased further by operating in vacuum.

One scheme for measuring the force gradients and the forces of a particular tip–sample separation over a spot on the sample is to drive the cantilever at ω_0 and to monitor the reduction in the vibration amplitude, $A(\omega_0, z)$, due to the shift in the resonance frequency of the system. Far away from the surface (beyond the interaction distance), the function $A(\omega, z)$ is Lorentzian:

$$A(\omega,\infty)=A_0\frac{\omega_0/\omega}{\left[1+Q^2\left(\frac{\omega}{\omega_0}-\frac{\omega_0}{\omega}\right)^2\right]^{1/2}} \quad (25)$$

When the tip interacts with the surface, the amplitude of the cantilever (being driven at ω_0) can be written

$$A(\omega_0,z)=A_0\frac{\omega_0'/\omega_0}{\left[1+Q^2\left(\frac{\omega_0}{\omega_0'}-\frac{\omega_0'}{\omega_0}\right)^2\right]^{1/2}} \quad (26)$$

This measured amplitude can be used to calculate the new resonance frequency, ω_0', which is directly related to the *force gradient* at the particular location above the sample by the expression

$$\frac{df}{dz}=m_{\text{eff}}^{1/2}(\omega_0^2-\omega_0'^2) \quad (27)$$

The force can be deduced by measuring df/dz as a function of z and then integrating:

$$f=\int\frac{df}{dz}\,dz \quad (28)$$

In practice, sensitivity can be improved by driving the cantilever at the resonance frequency at the point where the force gradient is maximized. The maximum sensitivity for the values of df/dz and f may be estimated assuming (1) some functional dependence of the force with respect to the separation and (2) the closeness of the steps of measurement. Estimates of typical microcantilever parameters yield a value for the highest force gradient sensitivity $df_{\text{min}}/dz = 3\times10^{-6}$ N/m. Considering an inverse square force with a tip–sample spacing of ~100 nm, the maximum force sensitivity would be $f_{\text{min}} \approx 2\times10^{-13}$ N.

Besides contact and non-contact imaging, the AFM can be used as a *nanoscale surface force apparatus* to measure adhesion between materials and the indentation/deformation properties of surfaces under extremely small loads (Burnham and Colton, 1989; Burnham et al., 1990). The atomic-scale basis of friction – *nanotribology* – can also be studied with a modified AFM by sliding a lightly loaded tip along the surface and measuring the lateral resistance to the motion. Various AFM applications are illustrated by several examples in the following discussion.

3.1 Atomic Force Imaging

The atomic force microscope can image both conducting and non-conducting materials. Sample conductivity is therefore not important, providing an advantage over the STM. In fact, the use of non-conducting tips allows studies of electrochemistry without interference from a conductive tip. In the *constant force mode*, the z-height of the sample is varied so as to keep the cantilever deflection, i.e., the repulsive force between tip and sample, constant. Without large scale surface deformations, this repulsive force may be viewed as occurring between hard spheres, which represent an atomic-sized tip asperity, and the sample surface atoms. This is a similar view to that taken for analysis of He scattering data.

In order to avoid tip-induced surface deformations, it is desirable to scan at the lowest possible forces. Many experiments address issues that do not require the cleanliness of an ultrahigh vacuum environment. Imaging in air, however, raises a difficulty regarding the minimum force with which to scan. This is due to the meniscus force of fluid contaminant layers, adsorbed water vapor or hydrocarbons, on the tip; such difficulties can be avoided by scanning with tip and sample immersed in a fluid.

The seminal work of Binnig et al. (1986) proved the feasibility of AFM by imaging aluminum oxide in air. With contact forces of the order of 10^{-8} N, a lateral resolution of ~ 30 Å was demonstrated although, at that point, they predicted the possibility of resolving ~ 1 Å corrugations with a period of > 1 Å. Atomic resolution was soon attained, and can now be achieved on a variety of materials using commercial instruments.

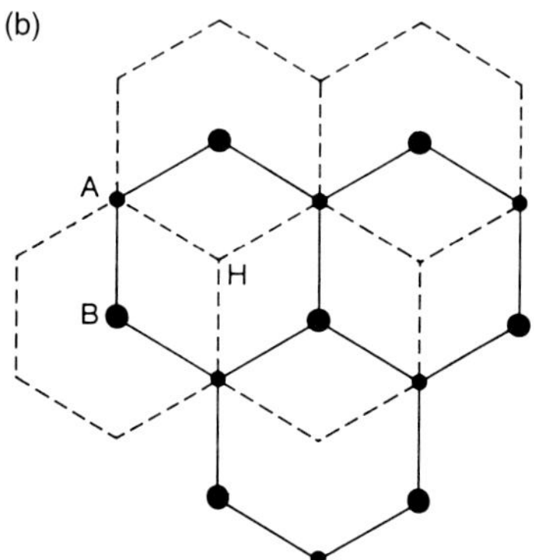

Figure 119. (a) Atomic resolution image of HOPG imaged in paraffin oil (Marti et al., 1987). (b) Diagram of the top layer of graphite illustrating atoms in "A" and "B" sites and the hollow (H) (Batra and Ciraci, 1988).

3.1.1 Graphite

AFM images of the surface of HOPG imaged in paraffin oil (Marti et al., 1987) are presented in Fig. 119 showing a centered hexagonal pattern. As for an STM analysis of graphite, the non-equivalence of the graphite surface atoms due to the lateral displacement must be considered, such that "A" atoms have atoms directly below while "B" atoms sit on top of hollows in the second layer. By performing self-consistent field calculations of the tip–surface interaction, including electronic relaxation, Batra and Ciraci (Batra and Ciraci, 1988) found that A and B sites are nearly equivalent in a typical AFM image. Lower imaging forces might actually image protrusions at hollow sites due to core–core repulsions producing the centered-hexagonal pattern.

3.1.2 Insulators

Cleaved LiF(100) surfaces provide a good illustration of the magnitude and nature of the corrugation measured by AFM (Meyer et al., 1991). An image of an LiF(100) terrace are shown in Fig. 120a; protrusions separated by ~ 2.8 Å and with a height of ~ 0.5 Å along [001] and ~ 0.3 Å along [011] are observed for a flat terrace. The corrugation amplitude was compared to data from helium scattering experiments, where hard sphere scattering was used. The ionic radius of Li^+ is 0.68 Å while that of F^- is 1.33 Å. Assuming that an oxygen atom of radius 1.5 Å sits at the apex of the SiO_2 cantilever, as depicted in Fig. 120b, the observed corrugations are in general agreement with a hard sphere model. In this picture, therefore, the F^- ion gives a negligible contribution to the corrugation so that the image almost entirely represents Li^+ ions at the surface.

3.1.3 Metals

The first atomic resolution images of gold atoms with AFM were performed on evaporated gold films on mica (Manne et al., 1990). This work demonstrated the differences in applied imaging forces when scanning in air against scanning in a liquid. Figure 121 illustrates 45 × 45 Å processed images of gold in air using a force of 10^{-7} N, and in water using a force of ~ 3×10^{-9} N. Both show that atomic resolution is attainable, but a higher imaging force is necessary in air to overcome the meniscus force on the tip arising from water and/or hydrocarbon contaminants on the surface.

To demonstrate this effect, attractive van der Waals and surface tension forces between tip and sample were quantitatively measured (Manne et al., 1990). They performed experiments to compare cantilever

(a)

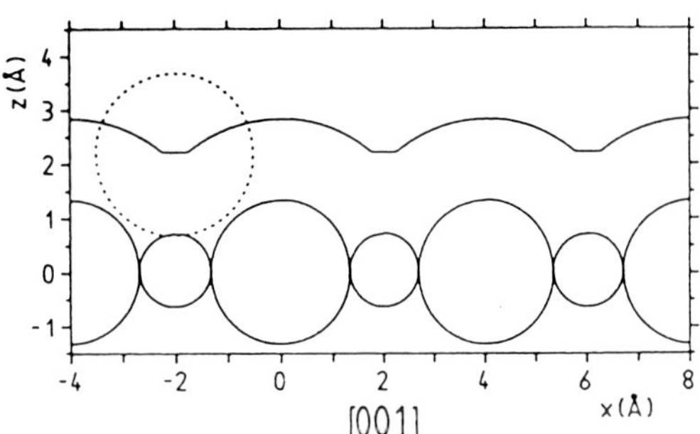

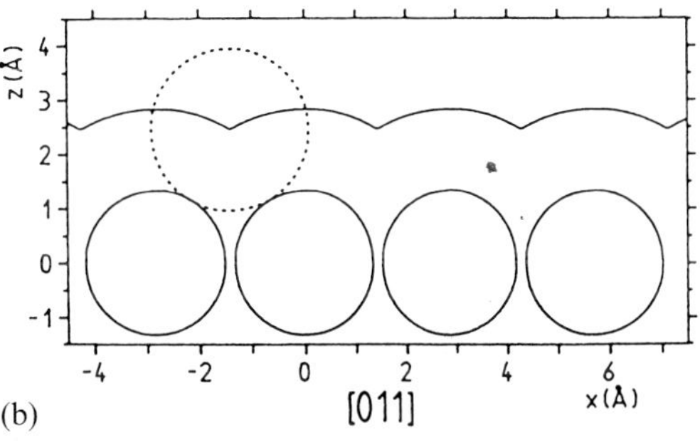

(b)

Figure 120. (a) 28 Å × 38 Å image of the LiF(100) surface; (b) simulation of scans along [001] and [011] using a contact hard sphere model between atomic protrusion on the tip (dotted sphere) and the surface, represented by spheres showing the ionic radii of Li and F (Meyer et al., 1991).

(a)

(b)

Figure 121. (a) Processed image of the surface of an evaporated gold film in air taken with a force of $\sim 10^{-7}$ N. (b) Processed image of the surface of an evaporated gold film in air taken with a force of $\sim 3 \times 10^{-9}$ N (Manne et al., 1990).

deflection as a function of vertical separation in air and water environments as the tip approached, touched, and was withdrawn from the surface of cleaved mica. In air, they found a significant hysteresis in this cycle and breakaway in the 10^{-7} N range, while in water the breakaway occurred in the 10^{-9} N range due to a reduction in the surface tension on the tip in the fluid environment.

The ability to image in various provides similar advantages as for STM. Different deflection responses to force gradients at liquid–air and the liquid–solid interfaces provide a means to make precise liquid film thickness measurements (Weisenhorn et al., 1991). Imaging at lowers in liquids is particularly advantageous for studying biological molecules adsorbed on various substrates. In addition, use of a non-conducting tip permits electrochemical reactions to be followed without influence from the tip, as in the case of STM.

3.1.4 Films

Imaging thin films using AFM provides a means to assess the structure and atomic order under various processing and growth conditions. Images of reactive films that have been removed from vacuum and exposed to air would be expected to show contamination. However, imaging in a reactive solution that removes the contaminants can provide structural information after chemical cleaning has been performed. AFM studies have been performed on thin Bi films deposited on cleaved mica substrates. Images of Bi films taken in a 0.01 M HCl solution exhibited areas in which the removed oxygen adlayer reveals a Bi(111) lattice, as shown in Fig. 122.

Buckyballs, or fullerenes (C_{60}), are a new class of materials that have sparked a good deal of interest recently. The entire range of their physical properties are under investigation, and the bulk and surface structural aspects of these materials are of particular importance. While the bulk structure had been reported to be face centered cubic (f.c.c.), the surface structure of thick fullerene films does not necessarily need to be a bulk truncation. In one study (Snyder et al., 1991), 1500 Å-thick fullerene films grown on $CaF_2(111)$ substrates in vacuum were imaged in air with ~ 10 nN force, as shown in Fig. 123. These studies corrobo-

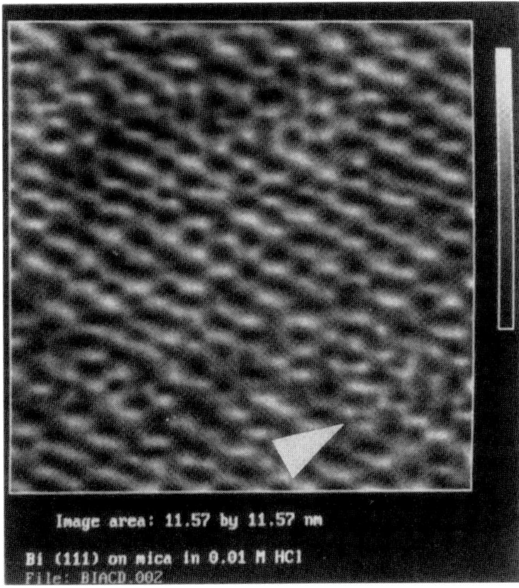

Figure 122. 116 Å × 116 Å image of a Bi(111) film on mica in 0.01 M aqueous solution of HCl (Weisenhorn et al., 1991).

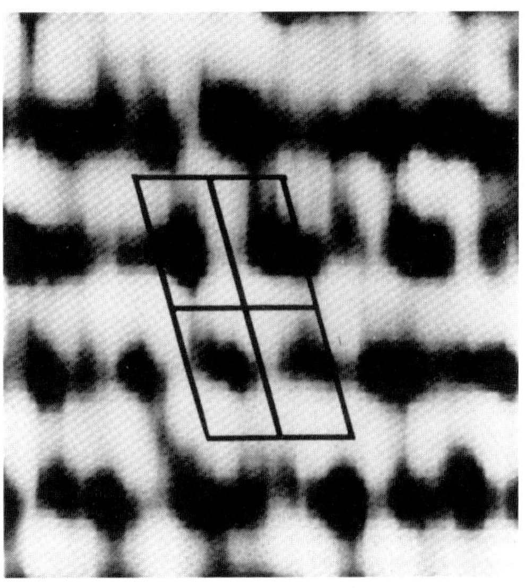

(a)

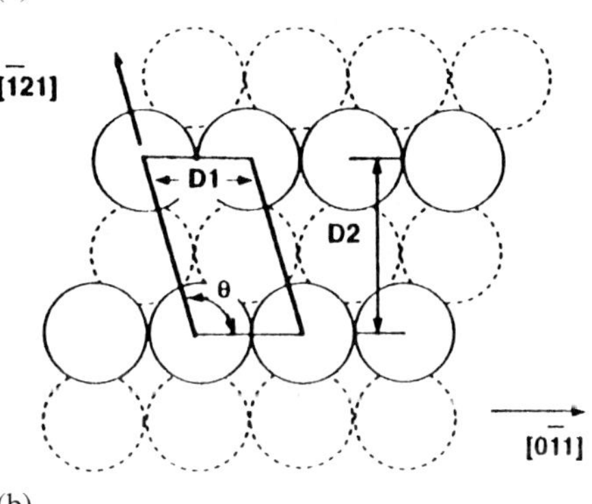

(b)

Figure 123. (a) Image of the surface of a fullerene thin film showing an exposed (311) plane. Four unit cells are traced on the image and the surface structure is traced (Snyder et al., 1991).

rated the fact that the material forms an f.c.c. structure. Interestingly, it was determined that the thermodynamically stable surface has a (311) orientation, not (111) as might be assumed on the grounds of minimization of surface energy. On the basis of these measurements, it was proposed that a gain in entropy by permitting free rotation of these unique molecules, as well as additional disorder allowed at an open surface, may lower the surface free energy of the (311) plane relative to the (111).

3.1.5 Polymer Surfaces and Metal Films on Polymer Substrates

Thin polymer films, particularly polyimides, have important applications for advanced microelectronics devices fabrication and packaging. Langmuir–Blodgett and spin-coating technique, as well as vapor phase deposition, may be employed to deposit monolayer and multilayer polymer films on various substrates. Issues which need to be addressed include deposition processes and curing, chemical bonding to the substrate surface, film stability and integrity, reactivity at the film surface, deposition and chemical bonding of metal layers on the film. AFM images establish film uniformity and morphology and the structure of metal adlayers. Recent work (Unertl et al., 1991) has shown that polyimide films are rather soft and the AFM tip interaction with the film is strong enough for the surface to be ploughed. Ploughing can be exploited to write lines with forces as low as 20 nN. As shown in Fig. 124, wide scan zero-force images on fresh polyimide foils exhibit fair uniformly with rms roughness in the 10–75 Å range. The image of an evaporated gold film on polyimide is particularly interesting. Evaporation of one monolayer increases the roughness due to non-uniform wetting of the foil and the formation of Au islands. In addition, Au reduces the ability of the tip to plough the surface. Further Au growth did not result in increased roughness, but smaller Au grains were observed as more of the surface was covered.

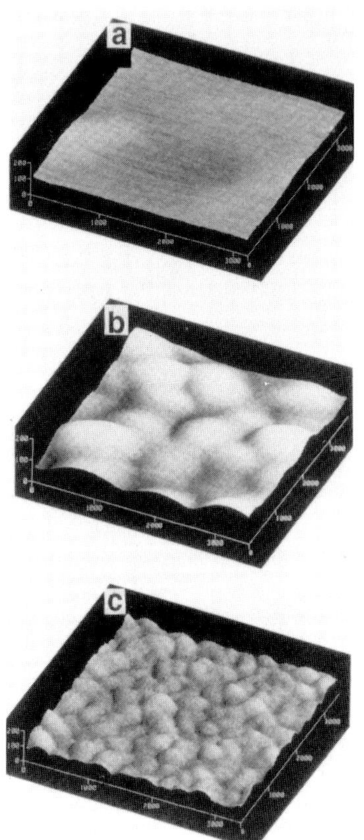

Figure 124. (a) AFM image of clean PMDA-ODA (poly[*N*,*N*′-bis(phenoxyphenyl)pyromellitimide]) film; (b) PMDA-ODA film coated with an ultrathin 3 Å gold film; (c) PMDA-ODA coated with a 50 Å thick gold film (Unertl et al., 1991).

3.1.6 Biological Molecules

Since the structural aspects of biological molecules are so important, it is useful to study both model organic layers and natural biological materials. In such studies, imaging in solution is useful for study under physiological conditions, as well as experimentally minimizing the imaging force to prevent tip-induced damage, e.g. ploughing or dragging of the material on the substrate.

Hydrated dimyristoyl-phosphatidylethanolamine (DMPE) bilayers have been studied (Zasadzinksi et al., 1991) as model systems to study structural aspects of biological membranes as a basis for the understanding of protein structure and organization. The DMPE layers were prepared by Langmuir–Blodgett techniques on mica substrates. The AFM image shown in Fig. 125 was taken at 1–10 nN and revealed structures separated by ~0.7 nm with protrusions along these structures every ~0.5 nm assigned as the headgroups of the molecules. Comparing the measured area/molecule (0.4 nm^2) with independent measurements proved that the images represented individual molecules. In addition, the imaging force was increased so as to “push through” the film so that mica could be independently imaged. Detailed analysis of the bilayers showed imperfect packing with several defects and vacancies. Repeated scans exhibited little damage, and demonstrated the occurrence of thermal motion in the fluid environment at room temperature.

Among the biological applications of the AFM, a major effort has been put into imaging biological macromolecules. As

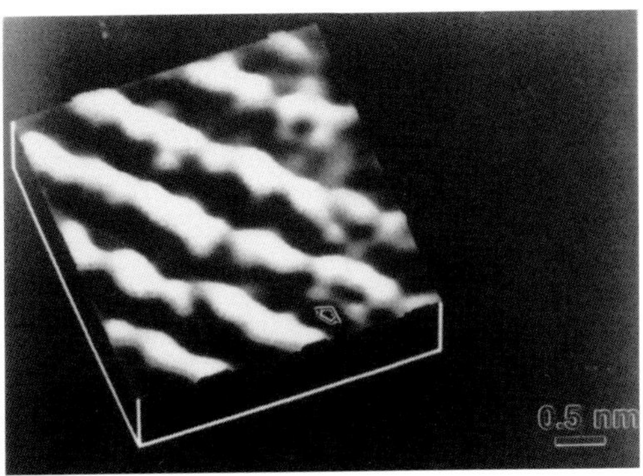

Figure 125. Image of the polar region of a DMPE bilayer deposited by Langmuir–Blodgett techniques. Rows are spaced by 0.7–0.9 nm. The modulation along the rows spaced every ~0.5 nm is from the individual DMPE molecule headgroups (Zasadzinski et al., 1991).

with STM applied to such systems, one of the problems is in properly anchoring macromolecules on the substrate so that they are not moved by the probe. Samples are prepared by covalent attachment to fatty-acid monolayers on mica or graphite supports. In the case of AFM, this is a major concern, and imaging is done in solutions which allow minimal interaction forces, such as water or ethanol. In addition, operating at low temperatures is envisioned as another means to reduce or prevent motion under the tip.

It has been estimated that sequencing of deoxyribonucleic acid (DNA) with an AFM could be performed ~100 times faster than with present technology, and that the technique could be applied in research focused on the binding of proteins, for example, at specific sites of the macromolecule (Hansma et al., 1991). Recent work in imaging and sequencing synthetically-produced DNA with large tagged bases at room temperature has indeed demonstrated the ability of the AFM to identify both the normal and tagging nucleotide bases (Hansma et al., 1991). Figure 126 shows a single-stranded 25-mer DNA image acquired under water. The long band at site 22 is the tag base (fluorescein) and, with higher magnification, bands 24 and 25 have been identified as thymine and adenine using molecular modeling of van der Waals profiles of the natural bases. While such work is in its infancy, improvements in sample preparation and instrumentation development are expected to satisfy both clinical and research goals in the future.

3.1.7 Adsorption Dynamics of Biological Molecules in Real Time

Repetitive scanning while organic molecules in solution interact with a surface can provide a real-time picture of adsorption dynamics and can provide evidence for mechanisms of self-assembly and ordering. In one study, the adsorption of immunoglobulin G [4-4-10 $IgF_2(\varkappa)$] – 18 μg/ml in a phosphate-buffered saline solution – onto a mica surface was studied. Approximate layer-by-layer growth occurred with each layer forming as a contin

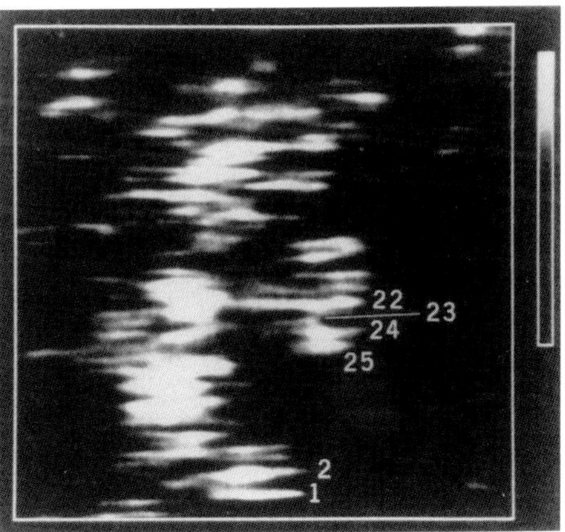

Figure 126. 9.5 × 9.5 nm image of a synthetic 25-mer DNA macromolecule imaged under water. The numbers identify various bases along the chain (Hansma et al., 1991).

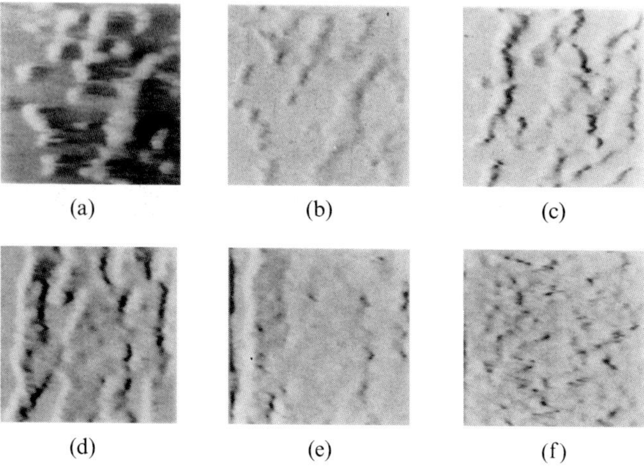

Figure 127. Sequence of images of immunoglobulin adsorption onto mica. Between (c) and (e) a monolayer is formed, after which time a second monolayer begins to form. The total elapsed time for this sequence was ~ 40 minutes (Lin et al., 1990).

uously growing aggregate, as seen in Fig. 127. Isolated molecules readily desorb, and therefore, lateral interactions are important to force the formation of a stable layer.

3.2 Nanoscale Surface Forces

The elasto-plastic behavior and hardness of a material is typically measured by its deformation response to an applied force. Microindentation probes to obtain this type of information have been employed for several years. Recognizing the need to probe structures with considerably smaller dimensions at increased force sensitivity, there has been an effort to further localize the area over which the measurement force is applied. The commercially-available nanoindenter, which can resolve forces with a sensitivity of 300 nN and 0.4 nm depth resolution, represents one step to satisfy these criteria.

The AFM provides orders-of-magnitude improvements over the nanoindenter, not only by superior performance in force and depth sensitivity (1 nN and 0.02 nm, respectively) for repulsive contact forces but also for use as an analog to the surface force apparatus (Israelachvili and Adams, 1978). Since both attractive and repulsive forces localized over nanometer-scale regions can be probed, forces due to *negative loading* of the probe from the van der Waals attraction between tip and sample prior to contact, or from *adhesive forces*, which occur subsequent to contact, can be investigated. The work of Burnham and Colton was the first to demonstrate the use of the AFM as a nanoindenter and to measure surface forces in contact with, and in proximity to, surfaces (Burnham and Colton, 1989; Burnham et al., 1990); the instrument depicted in Fig. 117a shows hard tungsten tips for indentation.

The contact behavior on indentation can be expressed by a loading curve, which is a plot of the relation between loading and penetration depth. In principle, the loading force is directly proportional to the cantilever deflection (ΔD). The penetration depth is ($\Delta z - \Delta D$), where Δz is the z-piezo displacement. The loading curves are expected to exhibit characteristics between two extremes – an ideally elastic material, where the loading and unloading follow the same curve, versus an ideally plastic material, where the material is irreversibly

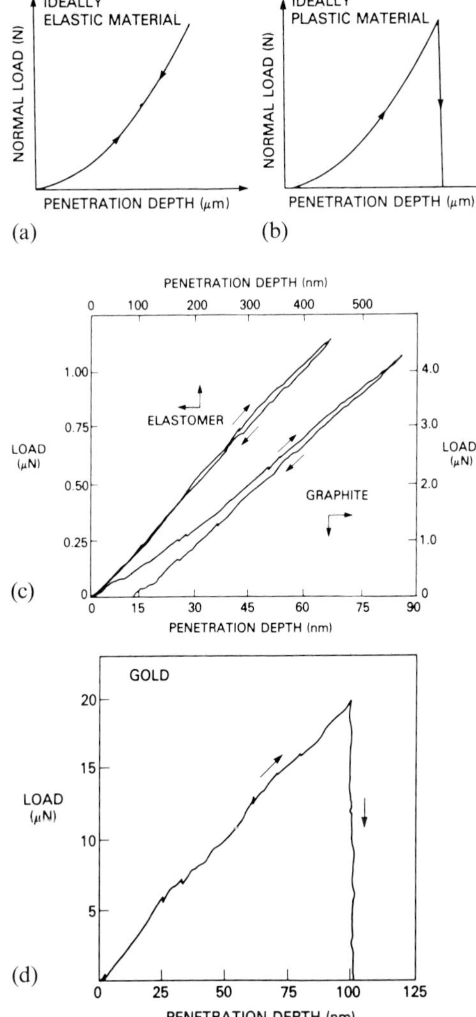

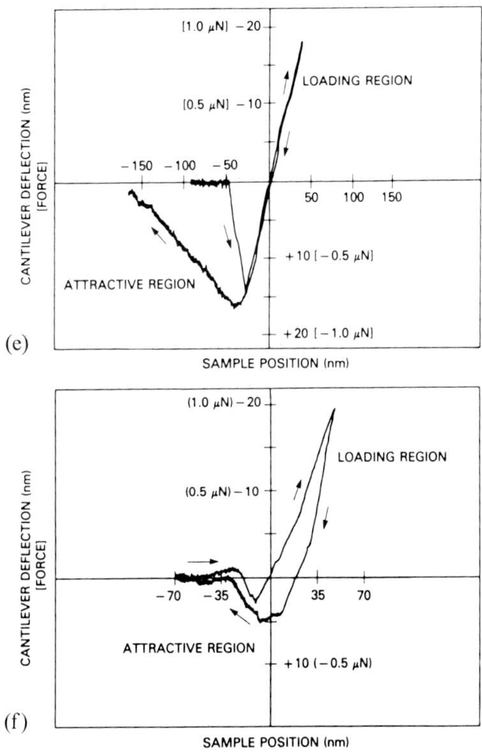

Figure 128. Expected loading curves for (a) ideally elastic and (b) ideally plastic materials; (c) experimental loading curve for an elastomer and for graphite showing elastic behavior; (d) experimental loading curve for a gold foil showing plastic behavior; (e) surface force interaction between a tungsten tip and graphite; (f) surface force interaction between a tungsten tip and gold foil (Burnham and Colton, 1989).

deformed, resulting in hysteresis between the loading and unloading curves. The contrast between elastic and plastic materials for nano-indentation of ≤ 100 nm can be seen in Fig. 128a, b. Figure 128c, d shows loading curves of an elastomer and of graphite, where an elastic-type response is observed, and a loading curve of gold, which exhibits a plastic response. The small hysteresis observed near zero loading for graphite is due to adhesive forces, as discussed below.

While these AFM nanoindentation studies involved forces on the μN scale and indentations on the 100 nm scale, an excellent example of the exploitation of AFM sensitivity is illustrated in Fig. 128e, f. Here the focus is upon measuring forces at extremely small loads, and the data is presented as force or cantilever deflection versus probe tip position. For a graphite surfaces, the tip jumps in towards the surface before contact indicating an attractive interaction of the order of ~ 100 nN between the tip and the surface due to the van der Waals interaction. The attractive force operates over ~1 nm until a repulsive force is felt. After contact, adhesion is indicated by the additional force required to break contact; in this case, the tip slides out, with the attractive force diminishing linearly. For gold, the tip slides in and out. A simple model has been developed to account for observations using various tip sizes and loadings, and for four possible types of behavior – jump in/slide out, jump in/jump out, slide in/jump out, and slide in/slide out. These processes reflect differences in the surface energy between graphite and gold. They may be modeled by considering the potential energy diagram of the entire measurement system – a cantilever spring potential obeying Hooke's law and a parameterized tip–surface interaction providing attraction and repulsion. A strong interaction based on van der Waals forces can cause the tip to jump in while a weaker interaction produces a sliding action. On the way out, deformation and adhesive forces required greater forces to remove the tip from the surface.

As we have noted, the role of surface forces has major implications of AFM and STM imaging; for example, elastic deformation has been used to explain the anomalous corrugations in graphite images. Furthermore, it has been estimated that surface forces produce pressures beneath the tip at zero applied load that are of the order of 4 MPa for PTFE (polytetrafluoroethylene) and 300 MPa for Al_2O_3. Such pressures are comparable to those involved in tribological testing. In addition, in a non-contact surface force image, topographic information must be independently analyzed.

3.3 Nanotribology

Nanotribology is a tropic of extreme interest since it addresses the microscopic origin of friction and the role of thin films as lubricants to reduce frictional forces between contacting objects. Observation of atomic scale features by measuring the lateral deflection of a loaded AFM tip being scanned across a graphite surface represents a first step in probing the origin of such microscopic frictional forces (Mate et al., 1987). In the instrumental configuration, interferometry was used for two orthogonal motions of the tip. The tip was tungsten wire bent at one end so that two orthogonal deflections could be detected. This arrangement has since been updated to a sensitive 2D capacitance probe (Neubauer et al., 1990) which can be used for nanotribology studies as well as for contact and non-contact imaging.

Figure 129 shows a plot of sidewards tip deflection (sidewards force) versus sample position for three different vertical loading forces. Focusing on a load of 2.4×10^{-5} N, it is evident that the frictional force has two components – a constant offset and an oscillatory part which varies with a periodicity of the order of the lattice spacing of graphite. The constant component can be viewed as the classical non-conservative frictional force. Thus the constant component varies nearly linearly with load. Thus the slope is the classical coefficient of friction, whereby energy is dissipated by generation of phonons or by other damping mechanisms. Deviations from linearity can be attributed to plastic deformation at the points of contact. The oscillatory part is particularly interesting since it represents a stick-slip motion of the tip as it is being scanned across the surface. This motion involves a conservative force due to elastic deformation associated with

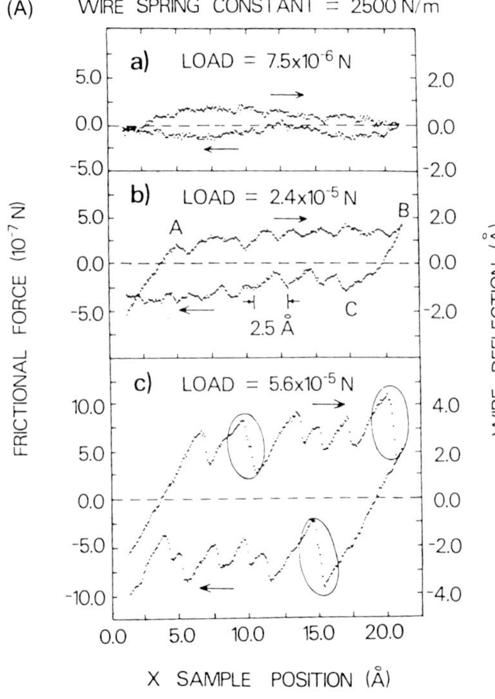

Figure 129. (A) Plot of frictional force versus sample position for a tungsten tip interaction with a graphite surface under three different loads; (B) frictional force in the x-direction as a function of x and y for a 5.6×10^{-5} N load (Mate et al., 1987).

the tip traversing a corrugated surface. During a scan under an applied load, the tip is repeatedly trapped in the periodic potential wells of the surface until the spring force equals the force trapping it. When these forces are equal, the tip slips towards the next corrugation as it relaxes. The observation of atomic-scale features is a surprising result since the tips had estimated radii of ~ 3000 Å and deformation is expected to be significant due to the large load. While atomic asperities at the tip surface reduce the effective contact area these are randomly distributed and would not be expected to produce periodic features. In order to explain these observations, it was proposed that the localized periodic force component associated with the interaction of tip asperities with the surface must be significantly larger that the non-conservative part, which scales with contact area. Furthermore, it is possible that the tip may be supported by a film which only allows for a single microtip to interact and provide the frictional force. A two-dimensional image of graphite which shows the frictional force (x-deflection) as a function of position is presented in Fig. 129 B.

3.4 Non-Contact Imaging

3.4.1 Van der Waals Forces

In the first demonstrations of non-contact imaging, profiles taken in the constant force gradient mode were reported for a grooved Si surface and for an Si surface partially coated with photoresist (Martin et al., 1987). The interaction due to the van der Waals attraction between the tip and the sample, separated by ~100 Å provides a means to profile the surface with a lateral resolution of > 100 Å. Figure 130a shows an image of grooves spaced ~2 μm apart on a Si wafer. It is important to note that the van der Waals interaction is material-dependent, so that contrast in constant

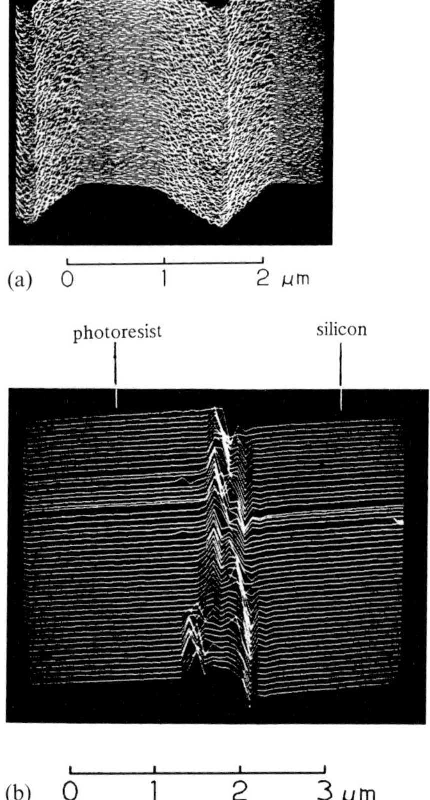

Figure 130. (a) Constant force gradient profile of a grooved silicon wafer; (b) constant force gradient profile of a flat Si wafer partially coated with photoresist (Martin et al., 1987).

force gradient images comes not only from surface topography but also if the tip begins to interact with another component on the surface. This is demonstrated in an image, of the junction between Si and photoresist, as shown in Figure 130b.

3.4.2 Electrostatic Forces

The measurement of the spatial distribution of charges on surfaces of insulating materials can be obtained by AFM (Terris et al., 1990). In this way, the microscopic spatial details of contact electrification, or triboelectrification (Stern et al., 1988; Terris et al., 1989; Terris et al., 1990), as well as ferroelectric domains (Saurenbach and Terris, 1990) at surfaces, can be obtained by imaging forces due to electrostatic or polarization charge, respectively. *Charge force microscopy* has demonstrated that the distribution of surface charge, deposited by inducing discharges between a grounded tip and sample (PMMA, sapphire, etc.), can be imaged with a lateral resolution of ≤ 0.2 μm with a vibrating tip situated 10–100 nm from the surface. The interaction force of interest comes from the interaction of the surface charge with an image charge induced in the tip. The tip also responds to the (uniform) charge distribution at the back of the sample due to varying its potential as well as to the van der Waals force. The van der Waals force attenuates as a function of sample–tip separation significantly faster than the electrostatic force, and can be neglected. A series of constant force gradient images take over a period of ~ 1 hour is presented in Fig. 131 A. At different sample biases, the image profiles of the surface charge change, as shown in Fig. 131 B, proving that the nature of the interaction is electrostatic and not due to surface topography. The sign of the charge can be obtained by taking such a series of images (Fig. 131 B), or by direct modulation of the sample bias (Terris et al., 1989). In Fig. 131 A, the charge is seen to decay and becomes spatially-broadened; charge redistribution over the surface is significantly more rapid than would be observed by a macroscopic measurement, indicating the unique information that spatial resolution provides. Furthermore, during contact electrification of polycarbonate and PMMA it was found that both signs of charge could be deposited simultaneously, as shown for polycarbonate in Fig. 132 (Terris et al., 1989). This novel observation has not been completely ex-

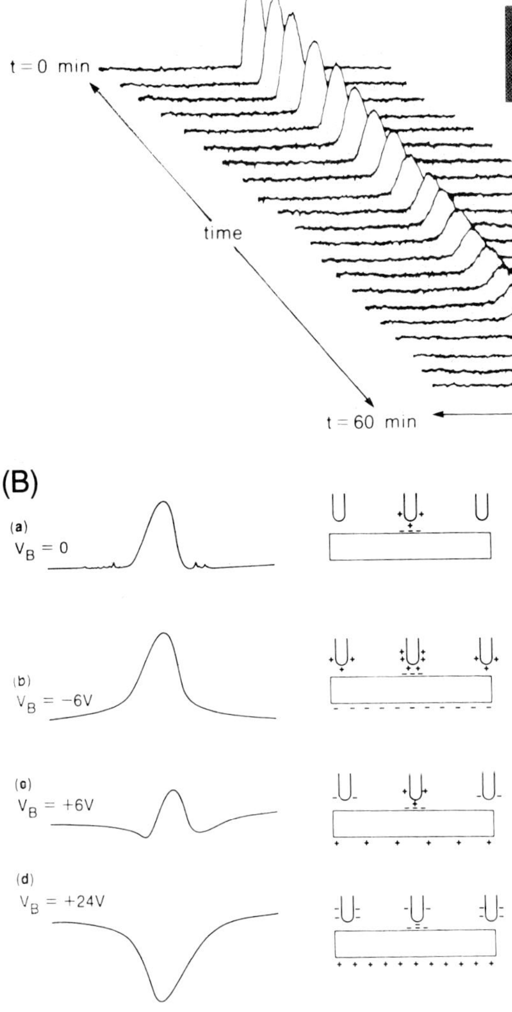

Figure 131. (A) Series of constant force gradient contours over a region of deposited charge shown in inset showing charge dissipation (Stern et al., 1988): (B) Constant force gradient contours for different bias voltages with charge distribution schemes (Terris et al., 1990).

plained and may be due to the method of charge transfer; however, the result suggest that charge measured by a macroscopic measurement may involve a balance between a bipolar distribution. The extreme sensitivity of this resonantly-enhanced technique has been estimated to be about three electronic charges, i.e., 3*e* (Terris et al., 1989).

3.4.3 Magnetic Forces

Magnetic microstructure is an important materials issue with applications extending from basic understanding of the interaction of microscopic magnetic domains to improving materials for high density information storage. The interaction of magnetized tips, either permanently or inductively magnetized, with a surface of a magnetic material can be imaged with the AFM operated in the non-contact detection mode. Both hard and soft magnetic materials can be imaged.

In the initial experiments to demonstrate the feasibility of the *magnetic force microscope*, images of thin film recording heads were obtained using two modes – running either AC or DC currents through the head to measure dynamic and static fields, respectively. With the magnetization modulated and with the tip set ~10 nm from the sample, pole pieces could be imaged with ~100 nm lateral resolution. To measure static fields, images with DC currents run in both directions were obtained

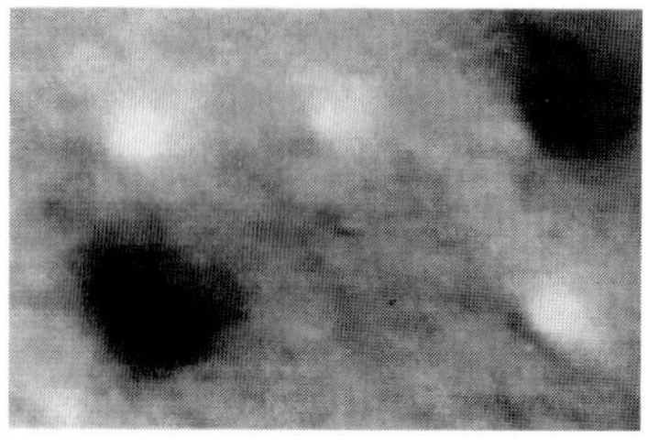

Figure 132. Images of five deposited charge regions on a polycarbonate surface (+, white; −, black) (Terris et al., 1989).

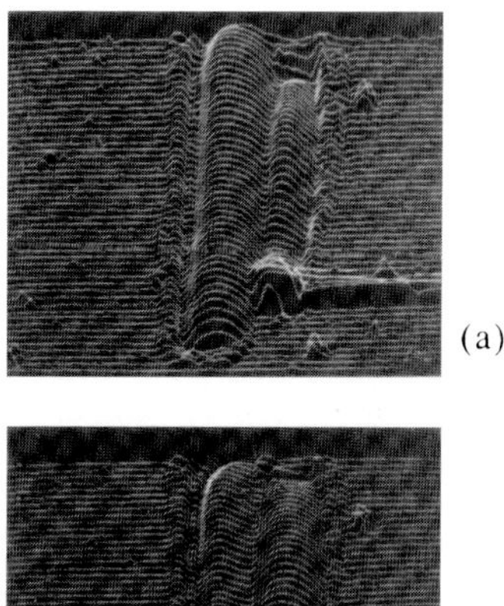

(a)

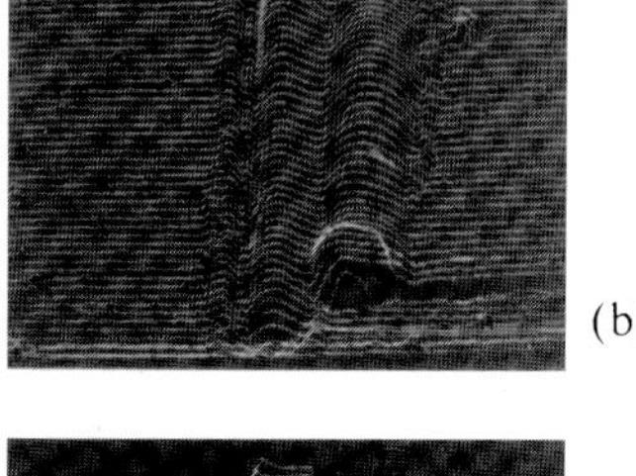

(b)

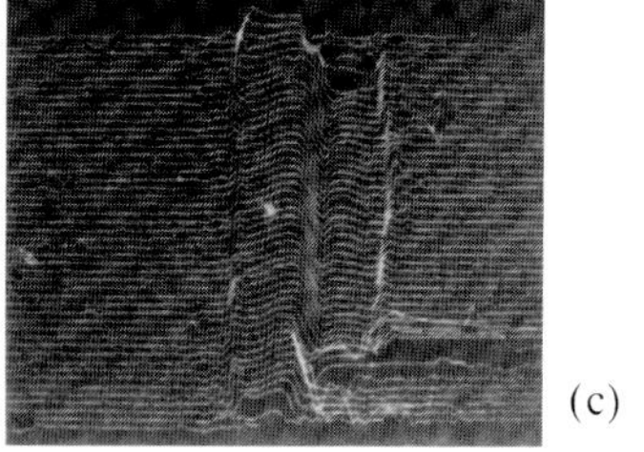

(c)

O 2 4 6 micron

Figure 133. Magnetic force images of a thin-film recording head with a current of (a) +10 mA, (b) −10 mA; (c) difference between (a) and (b) (Martin and Wickramasinghe, 1987).

(Fig. 133 a, b). While a net attractive interaction resulted due to tip-induced magnetization, for one current direction the left pole produced the main attractive force while for the other direction, the right pole dominated. A difference image (Fig. 133 c) shows a magnetic field pattern that defines the pole pieces.

Magnetic force images of single-crystal iron whiskers (Göddenhenrich et al., 1990 b) and permalloy thin films (Mamin et al., 1989) were obtained to demonstrate the ability of the technique to distinguish the structure of domain walls. In a constant force gradient image of the 2.5 μm permalloy thin film shown in Fig. 134, Bloch walls appear as bright or dark lines and the domain structure is comparable to that by Kerr imaging techniques. The shift from bright to dark at the arrow indicates the position of a possible Bloch line. These images were obtained with a bias on the sample to ensure a net attractive force and a magnetic tip at a distance of ~150 nm from the surface. When the tip was brought close to the surface of a 30 nm permalloy thin film, images became distorted, sug-

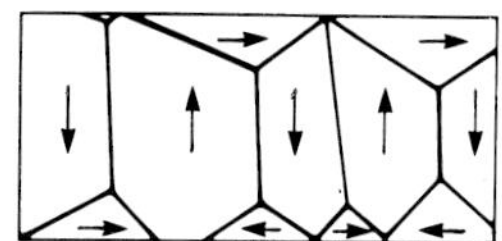

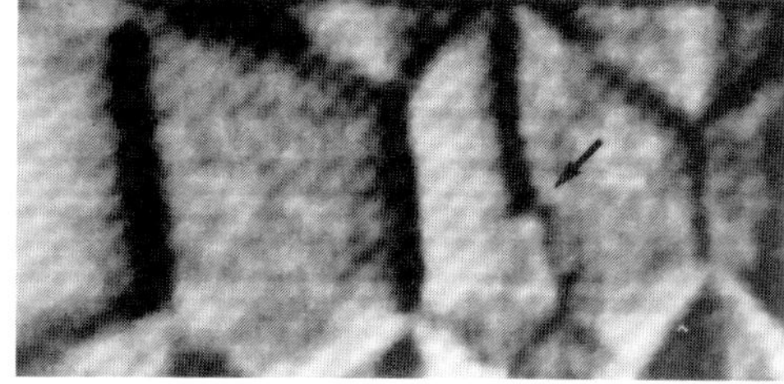

Figure 134. Magnetic field image of domain walls on permalloy thin film. The bright/dark lines signify Bloch walls and the arrow defines a possible Bloch line (Mamin et al., 1989).

gesting wall motion due to a reversible tip-induced magnetic structure in the sample when the sample coercivity is locally exceeded.

4 Manipulation of Atoms and Atom Clusters on the Nanoscale

A most challenging current goal of materials science is to manipulate atoms at the atomic level in order to fabricate high-density structures for technological applications ranging from microelectronics to bioengineering. The limit in the miniaturization of high density information storage devices or lithography is the development of the technology to produce and utilize structures of atomic dimensions. At this level, the STM and AFM can be viewed as read/write devices – that is, information can be read by scanning the surface with atomic resolution, manipulated by moving atoms or changing the states of atoms on the magnetic with the tip, and re-read by scanning over the modified surface. It is well-known that some types of surface modification can occur by uncontrolled tip crashes during STM operation. Controlled writing of grooves in gold through mechanical deformation was first demonstrated by Abraham et al. (1986).

Control of *atomic-scale* surface modification derives from the exploitation of the mechanisms involved in the tip–atom interaction and demands on ability to index the tip so as to successfully return to the same point on the surface repeatedly. In this context, a nanoscale memory element could be envisioned as the deposition, removal, or transfer of an adsorbate atom at a specific surface site or region using the STM tip (Quate, 1991). The information contained in this and other atomic-sized "data bits" could then be read with the same tip in the scanning mode. The prospect of actually realizing this vision has been demonstrated in recent STM studies, in which the xenon atoms were moved into desired locations and an "atom-matrix" pattern was subsequently imaged in ultra-high vacuum at liquid helium temperature (Eigler and Schweizer, 1990). It this useful to show that direct and indirect lithography ot atomic-sized structures can be achieved on electronic materials. In this regard, it has been shown that silicon atoms and atom clusters can be transferred between one location and another on the substrate using the STM tip as a carrier (Lyo and Avouris, 1991). This was achieved by pulsing the voltage on the STM tip to pickup and subsequently deposit the atoms. Besides bi-directional transfer of material between the surface and the tip (Lyo and Avouris, 1991), the attraction of chemisorbed atoms along a surface by the tip and tip-induced surface chemical modification has also been demonstrated. Extensions of these techniques to custom-form or manipulate species at the molecular level are the basis for site-specific chemistry and biochemistry.

The magnitude of forces involved for moving atoms along the surface over potential corrugations should be quite different from those involved in transferring atoms between the tip and the sample, or vice-versa (Eigler and Schweizer, 1990). Thus, the mechanisms involved for atom or cluster manipulation at surfaces and specific application of such structures require new physical ideas regarding magnitudes and distance-dependences of highly localized forces. Besides this work, surface modification by direct contact of tip and sample using a force microscope may be viewed as another new area for exploration and exploitation (Weisenhorn et al., 1990). Developing such capabilities opens up many possibilities for future applications in advanced materials technology.

4.1 Transfer of Atoms and Atom Clusters Between Tip and Sample

The first example of atomic-scale surface modification was reported by Becker et al. for the creation of structures such as the Ge(111) surface (Becker et al., 1990). The surface was modified by holding the tip over a surface site, raising the tip–sample bias to −4.0 V (20 pA current), and rapidly withdrawing the tip by ~1 Å. Following this procedure, images of the surface displayed isolated protrusions about 8 Å wide and ~1 Å high, as seen in Fig. 135. Since the process occurred with a higher success rate when the tip had previously contacted the surface, it was proposed that the Ge atoms, acquired from previous tip–surface contact, were being deposited from the tip onto the surface.

Positioning an STM tip in close proximity to the surface lowers the potential barrier to field-evaporation and this effect can indeed serve as a means to transfer atoms or clusters of atoms between the tip and the sample. The condition to confirm transfer via a field evaporation mechanism is the existence of a threshold bias voltage for deposition or removal of matter. Subsequent to the work of Becker et al. (1990), the existence of such a threshold for deposition at a selected site by pulsing the tip–sample bias was demonstrated (Mamin et al., 1990; Lyo and Avouris, 1991).

Mamin et al. (1990) utilized a field evaporation process to deposit ~10 nm diameter gold dots of height ~2.5 nm onto a gold surface using a gold tip as the source, as shown in Fig. 136. Such dimensions are significantly smaller than the 100 nm Ga lines which had previously been drawn by ion emission from a liquid metal ion source (Bell et al., 1988). The experiments of Mamin et al. have demonstrated the existence of a pulse height threshold below which the probability of deposition was nearly zero and above which the probability increased to nearly one. Examination of the dependence of the threshold voltage on the tip–sample distance confirmed a constant gap impedance indicating the need for a critical field of ~0.4 V/Å. Considering a model in which deposition occurs via ionization followed by field evaporation for an isolated tip, fields of 2–7 V/Å were predicted, and, for gold, values of ~3.5 V/Å have been measured. The critical field is lowered with the two electrodes in close proximity so that the atomic potentials begin to overlap and the effective barrier height for evaporation is lowered. Even under these conditions, estimates of 1–3 V/Å are still higher than the observed threshold

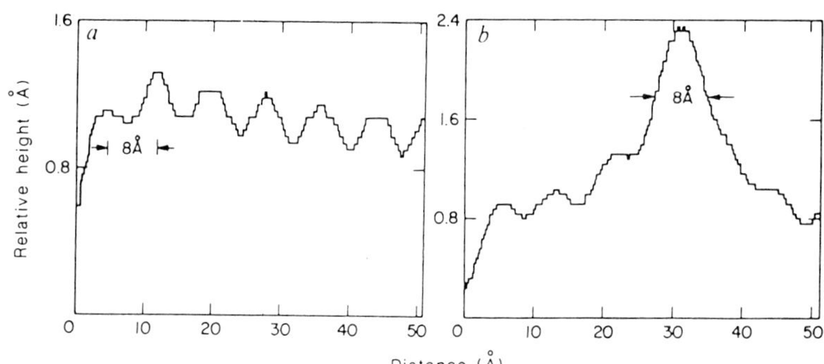

Figure 135. Scans of surface height (left) before and (right) after impression (Becker et al., 1987).

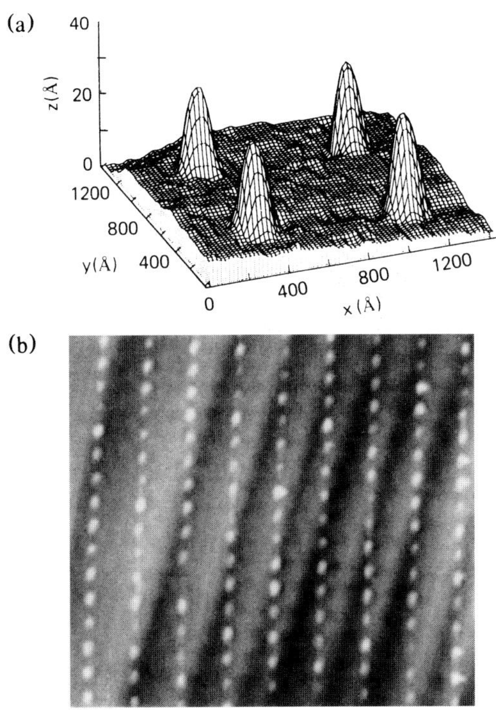

Figure 136. (a) Mounds of gold made by applying voltage pulses between a gold tip and the sample. (b) Array of mounds formed on a stepped gold surface (Mamin et al., 1990).

fields. While quantitative disagreements still exist, the presence of a bias threshold provides evidence that a field evaporation mechanism drives the transfer of material.

Not only can field evaporation occur from the tip to the sample. Transfer of material sample-to-tip through a field evaporation process should also be possible, and this effect has recently been demonstrated. In particular, the manipulation of Si atoms and Si clusters on an Si(111) surface has been demonstrated by Lyo and Avouris (Lyo and Avouris, 1991), whereby material was transferred in a controlled fashion from the substrate onto the tip for re-deposition at another place on the surface. Removal of Si atoms or clusters involved field-evaporation from the surface to the tip by a +3 V pulse after the tip was placed ~3 Å from the surface. It was found that smaller amounts of Si were removed when the tip was closer to the surface and the amplitude of the voltage pulse was made smaller. For example, Fig. 137 shows the removal and re-deposition of a single Si atom; here the tip–sample distance was ~5 Å and the pulse was +1 V. Based on current versus tip displacement curves, it was proposed that the Si most likely resides at the apex of the tip during transfer. In assessing the mechanisms for the transfer of material, especially at very small distances, consideration must also be given to the modification of the semiconductor covalent bonds by large local current densities, to chemical interactions between the Si atom(s) and the tungsten tip, and to mechanical effects. It should be noted that, while mechanical contact does disturb the surface, it cannot be used as the pulsed-field process for controlled material transfer.

4.2 Tip-Induced Lateral Motion of Atoms on Surfaces

The potential barrier to lateral movement of atoms is generally lower than the barrier for vertical motion, or removal. In order to use the STM tip to move an atom

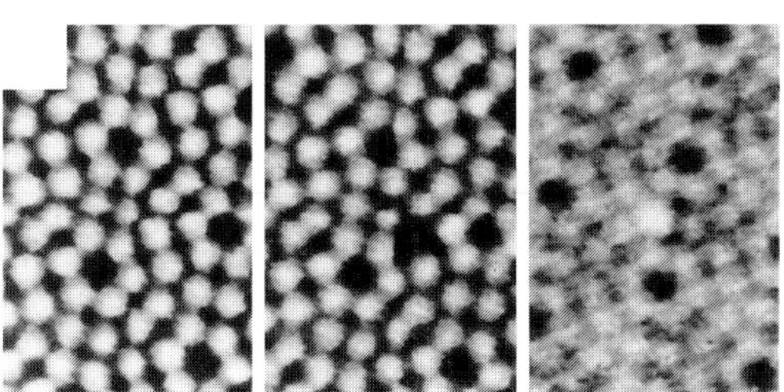

Figure 137. Removal and re-deposition of a single Si atom on an Si(111) surface, Left: before voltage pulse. Middle: after voltage pulse. Right: after re-deposition (Lyo and Avouris, 1991).

across a surface, two types of tip-atom interactions may be envisioned – van der Waals or electrostatic forces. Recently, Eigler and Schweizer demonstrated the precise position of xenon atoms on an Ni(110) surface and conducted preliminary tests regarding the lateral attractive interaction between the tip and the atom, as shown in Fig. 138 (Eigler and Schweizer, 1990). Xenon bonds to the surface by physisorption so the experiments must be performed at low temperatures. In the experiments of Eigler and Schweizer, the STM and the vacuum chamber in which it was housed were cooled directly with liquid helium. The advantage of this arrangement is that the system in thermal equilibrium is very stable and cryopumping by all the cold surfaces creates such a high vacuum that the sample surface can remain clean for weeks. The process of atomic positioning of Xe by lateral tip–atom interactions involved the following steps: (1) the Ni surface was exposed to Xe, (2) the tip was placed directly over a physisorbed Xe atom and lowered to increase the attractive force by increasing the tunneling current, (3) the tip/atom combination was moved to the desired site, (4) the tip was withdrawn by reducing the tunneling current and subsequent imaging was performed to establish the success of the operation. It has already been noted that the ability to image a closed-shell atom such as Xe at such small biases only exists because the Xe 6s resonance occurs near the Fermi level of the substrate (Eigler et al., 1991 b). In these experiments, both imaging and motion could be performed at a tip bias of 10 mV, while higher biases increased the interaction distance. The ability to displace Xe atoms was found to be polarity-independent, indicating that the van der Waals attractive interaction was the operative mechanism for atomic manipulation. It was also found that the close-packing distance between Xe atoms was fixed, being a property of the closed-shell atom, so artificial structures of higher density could not be obtained.

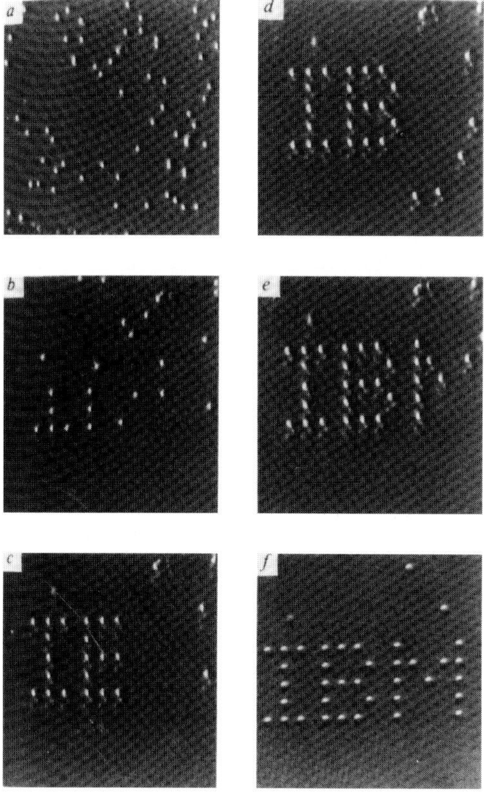

Figure 138. Sequence of STM images acquired during the construction of a patterned array of xenon atoms on an Ni(110) surface (Eigler and Schweizer, 1990).

Studying physisorbed atoms at low temperatures has provided a good deal of fundamental information on mechanisms for atomic manipulation. However, operating at these temperatures is experimentally restrictive. Atomic manipulation of chemisorbed systems at room temperature is a desirable extension of this work. Due to the stronger interactions of chemisorbed atoms with the surface, local manipulation may permit the formation of artificial ordered structures more readily since the atoms need not reach thermal equilibrium, e.g. by annealing, as is usually the case. Whitman et al. (1991 c) demonstrated that, by pulsing an STM tip set above sub-monolayer coverages of Cs atoms adsorbed on a GaAs(110) surface, the Cs atoms can be rearranged laterally through field-induced diffusion. The Cs/GaAs(110) system is particularly interesting since it has been discovered that, at such low coverages, Cs forms various one-dimensional structures along the $[1\bar{1}0]$ direction (First et al., 1989 a) and other two-dimensional structures at higher coverages (Whitman et al., 1991 a). The fact that one-dimension equilibrium structures form is consistent with the anisotropic GaAs(110) surface structure. With a distribution of Cs atoms on the surface, moderate pulsing of the tip bias was found to attracts Cs atoms towards the tip anisotropically so that new chains formed in the local region of the tip, as shown in Fig. 139. Larger pulses aggregated Cs into mounds around the tip. Other pulsing conditions produced ordered structures not seen on the Cs/GaAs(110) interface in thermal-equilibrium. The fact that pulse width and pulse amplitude result in larger effects in the tip region implies a field-induced diffusion process. The spatially-varying field produced by the tip produces a potential gradient, or a force, laterally and vertically with the attractive interaction towards the tip. Estimates predict that the polarizability of a Cs atom on the semiconducting surface is considerably larger than on a metal surface. However, considering the smaller surface potential corrugations on metal surfaces, field-induced diffusion might also be applicable for these systems.

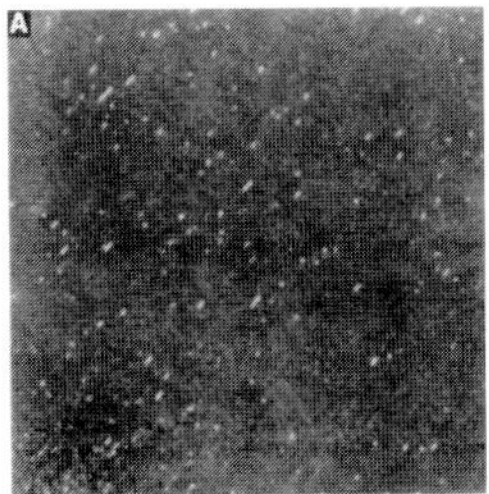

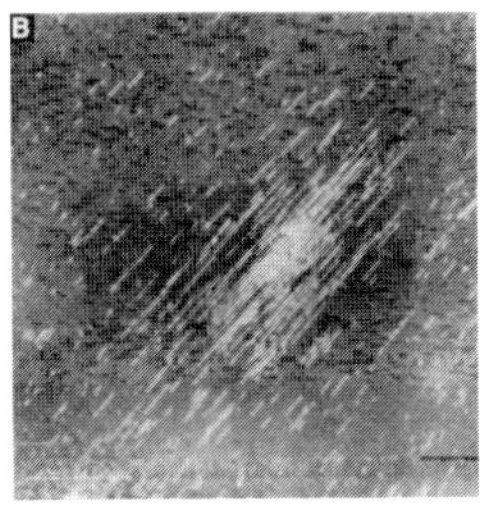

Figure 139. STM images of Cs on a GaAs(110) surface (A) before and (B) after pulsing the sample bias with the tip in the center of the imaged area (Whitman et al., 1991 c).

4.3 Nanoscale Modification by Tip-Induced Local Electron-Stimulated Desorption

Nanoscale modification of the chemically-passivated Si(111):H-1×1 surface in vacuo (Becker et al., 1990) and in air (Dagata et al., 1990) has been demonstrated recently. It was previously established that the Si(111) surface can be made entirely unreactive by saturating dangling bonds with a layer of atomic hydrogen deposited by an ex situ, in-solution chemical process (Higashi et al., 1990). Arsenic-passivation of Si(111) and Ge(111), performed in vacuo, had been demonstrated even earlier (Bringans et al., 1985; Olmstead et al., 1986). The H-passivated surface is one of the most ideal flat surfaces ever observed (Higashi et al., 1990). Becker et al. (Becker et al., 1990) have convincingly shown that using the STM in a field emission mode converts a localized area under the tip to a clean Si(111)-2×1 surface by the electron-stimulated desorption of H. STS measurements of the passivated surface regions showed no surface states, and spectroscopy of the 2×1 regions produced the previously-established Si(111)-2×1 surface state spectra (Stroscio et al., 1986). This result is significant in two regards. First, it shows that a low activation-barrier exists between the 1×1:H surface and the 2×1 π-bonded chain structure. Second, the localized field-emission-induced desorption process exhibits an energy dependence suggesting that H-desorption proceeds via excitation to the unoccupied σ^* level (~ 8 eV kinetic energy) with an efficiency of $\sim 3 \times 10^{-8}$. The spatial resolution depends on the geometry of the set-up, i.e. on the area of incidence of the field-emitted electrons. With the tip ~ 1000 Å above the surface and a bias of ~ 60 eV, the 2×1 region was found to be ~ 8000 Å wide. If the tip was ~ 40 Å away and had a lower bias, widths of ~ 40 Å were obtained. For such a system, subsequent decoration of the clean areas and possible selective etching might be envisioned to produce permanent nanoscale patterns.

Modification of the H-passivated Si(111) surface in air has also been reported (Dagata et al., 1990). The observed nanostructures appeared as 1–20 nm deep depressions with widths of the order of 100–200 nm. These are areas where oxidation has taken place, so oxygen occupies the de-passivated regions of the substrate. The tip-induced process was considered to proceed by field-induced ionization of oxygen, resulting in diffusion-enhanced oxidation. In light of the results of Becker et al. (Becker et al., 1990), field-emission induced removal of H should also be considered.

4.4 Nanoscale Chemical Modification

Various attempts at STM-assisted deposition or modification of chemical species at surfaces include etching lines on GaAs in an electrochemical cell (Lin et al., 1987), pinning large organic molecules to graphite surfaces by pulsing the tip bias (Foster et al., 1988), decomposing an organometallic (dimethyl cadmium) under an STM tip to write metal lines (Silver et al., 1987), and exposing polymethylmethacrylate resist using the localized tunneling current (McCord and Pease, 1987). In each of these examples, the presence of the tip created a reaction of the chemical species to obtain desired surface features.

4.5 High-Temperature Nanofabrication

Attempts at nanofabrication on Si surfaces maintained at high temperatures have recently demonstrated that the electrostatic tip–surface interaction coupled with the high diffusivity can permit atomic control, so that writing speeds of 10 nm/s can be obtained (Iwatsuki et al., 1992). High temperature imaging of Si surfaces has indicated that diffusion of the adatoms is high enough to form a 1×1 structure. For nanofabrication under similar conditions, the tip bias is raised and a constant tip–sample distance is maintained as the tip is scanned. The electric field strength is determined, not only by distance, but by bias, tip geometry, and tunneling current. Patterns with linewidths of 1–2 nm and depths of 0.3 nm have been obtained. Besides atom manipulation, larger structures can be formed. For example, applying a +4 V bias on the tip, the tip-induced electric field causes atoms to migrate to the tip and hexagonal pyramidal structures and quadrangular pyramidal structures have been formed on Si(111) and Si(100), respectively, with a sample temperature of approximately 600 °C, as illustrated in Fig. 140. When the polarity was reversed, craters were formed. The edges of these features revealed higher index planes or step structures.

4.6 Nanoscale Surface Modification Using the AFM

Manipulation of molecules on surfaces has also been demonstrated using the AFM. In particular, Weisenhorn et al. (Weisenhorn et al., 1990) used an AFM tip to write on zeolite surfaces by moving

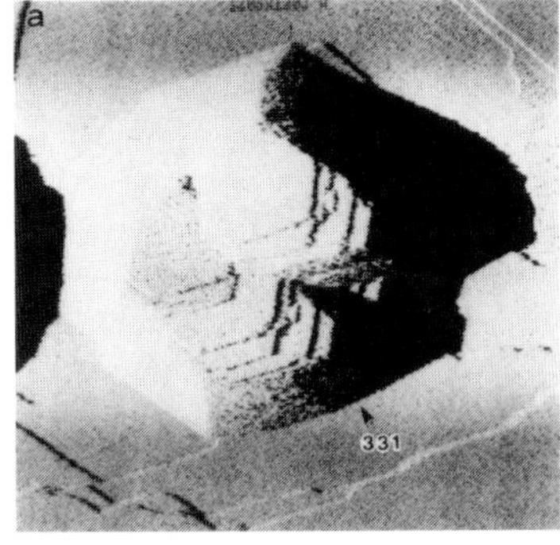

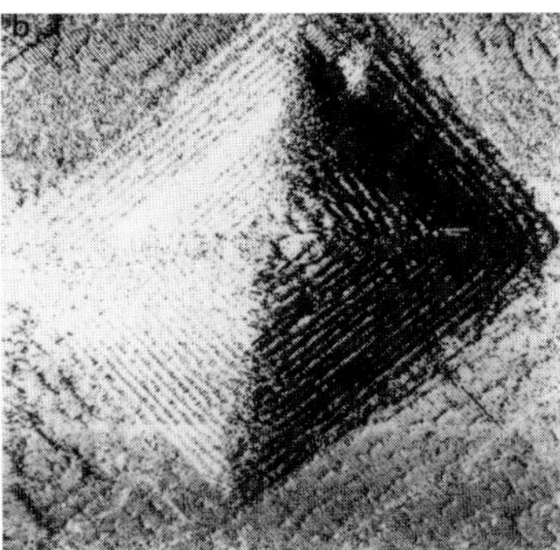

Figure 140. (a) Hexagonal pyramidal structures formed on Si(111), and (b) quadrangular pyramidal structures formed on Si(100) obtained by scanning the tip with a +4 eV bias over a sample maintained at ~600 °C (Iwatsuki et al., 1992).

molecules when large contact forces were applied. In this work, a layer of *tert*-butyl ammonium cations was adsorbed on a clinoptilolite surface in order to assist this. These cations are bound to the surface via surface ion adsorption and form clusters after injection of the solution on a time scale that could be monitored by the AFM. Using a force $>10^{-8}$ N at selected locations while scanning, an "X" was written into the layer. Subsequent imaging at $<10^{-9}$ N indicated that the clusters we connected in the pattern and that the pattern remained stable.

4.7 Towards Nanoscale Devices

While the experiments of Eigler and Schweizer (1990) attained wide notoriety, particularly, when an "atom-matrix" display of the IBM logo using Xe atoms was fabricated, the use of the Xe–Ni(110) systems as an "atomic switch" was also demonstrated (Eigler et al., 1991 a). In this regard, individual Xe atoms were reversibly transferred between the tip and the sample using voltage pulses of opposite polarities (± 0.8 V) at low temperatures. The state of the switch was subsequently queried by measuring the tunneling conductance, which changes if the Xe atom is bound to the tungsten tip or the nickel surface. This process is illustrated in Fig. 141. The change of state and conductivity measurement are performed with the tip fixed above the surface, so imaging the atom is not required to test the state of the switch. That the Xe atom has actually moved to the tip was inferred, however, from the absence of Xe in a subsequent image. The change in conductance is due to the different localized sites for the atom in its two possible positions, which also changes the

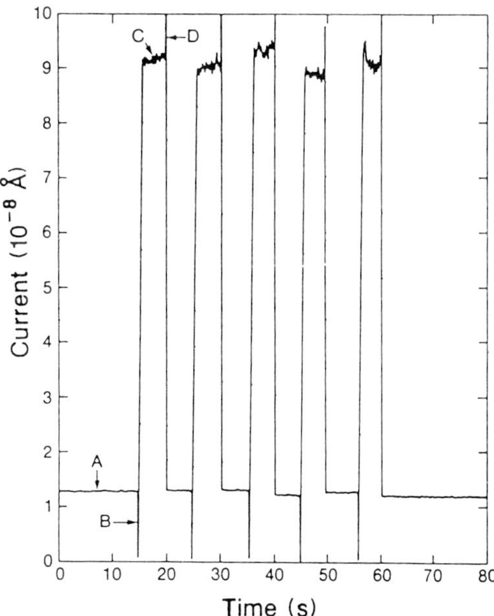

Figure 141. Time dependence of current through the switch during operation of the atomic switch (Eigler et al., 1991 a).

effective tunneling distance on which the current is exponentially-dependent. In contrast to the work discussed above, field-evaporation was excluded as a possible atom transfer mechanism since the ion states exist far above the Fermi level and no threshold electric field was required for the transfer. Two other proposed mechanisms still remain to be tested – electromigration and impurity heating, both due to the tunneling current in the vicinity of the Xe atoms.

5 Spin-Offs of STM – Non-Contact Nanoscale Probes

The STM and AFM techniques have been discussed in previous sections of this article. Their application to a wide variety of problems in science and technology have been realized over a relatively short period of time. The physics of the electron tunneling process provides the STM with capabilities for both topographic and spectroscopic imaging of surfaces on the atomic scale. In contact and non-contact modes, the AFM probes surface forces with extremely high spatial resolution and force sensitivity. STM and AFM can operate in vacuum as well a in air and liquids opening these surface techniques to a host of other problems not typically associated with surface science, notably the study of biomolecular systems.

STM and AFM are based on imaging in the near-field, i.e. the probe is in close proximity to the sample. Several new techniques that borrow from the philosophy and technology of STM/AFM instrumentation, i.e. adopting the near-field imaging process, using piezoelectric actuators, and, in some cases, taking advantage of spectroscopic capabilities, have either been demonstrated or proposed. Extensions of STM/AFM include probing the optical, thermal, electrical, or chemical properties of surfaces. In Table 2, various techniques – STM, AFM, and spin-offs – and their attributes are listed. Details of the techniques and their applications are discussed in the remainder of this section.

5.1 Scanning Near-Field Optical Microscope (SNOM)

The spatial resolution of standard optical microscopy, i.e. far-field optical microscopy, is diffraction-limited to lengths of the order of the wavelength λ (actually $\lambda/2.3$). Thus the resolution limit for imaging at optical wavelengths is ~ 0.2 μm. In the familiar far-field regime, an image is reconstructed all at once by a focusing lens. Significantly, optical imaging with considerably higher spatial resolution has been demonstrated using *near-field techniques*. Considering the STM as a near-field probe with the tip localized near the sampling region, sequential responses to the tunneling of electrons at discrete points over the scanned region are assembled to form the image. For a near-field technique, the source or the probe must be small; the spatial resolution is roughly limited by the aperture size or the probe-sample distance.

Near-field optical microscopy operates on the same principle as a physician's stethoscope where the detector is much smaller than an acoustic wavelength (Pohl et al., 1984). The optical analog is the scanning near-field optical microscope (SNOM) in which light is channeled down a metal-coated quartz tip with an ~ 10 nm aperture at its end, and the tip is scanned over the surface. In the first manifestation of this instrument (Pohl et al., 1984), the tip was sequentially moved and lowered into contact with the sample at each sampling point using piezoelectric drives. Using

Table 2. Near-field surface characterization probes.

Instrument	Acronym	Measurement parameter	Lateral resolution
Scanning tunneling microscope	STM	electron tunneling current	< 0.2 nm
Atomic force microscope	AFM	surface force/force gradient	< 0.2 nm
Tribological force microscope		lateral force	< 0.2 nm
Attractive force microscope		van der Waals force	< 100 nm
Charge force microscope	CFM	electrostatic force	< 200 nm
Magnetic force microscope	MFM	magnetic force	< 100 nm
Scanning near-field optical microscope	SNOM	near-field optical reflection	<25 nm
Photon scanning tunneling microscope	PSTM	near-field optical transmission	$\lambda/10$
Scanning thermal profiler	STP	surface temperature	< 30 nm
Scanning chemical potential microscope	SCPM	thermoelectric voltage	< 1 nm
Optical absorption microscope	OAM	effects of light absorption	< 1 nm
Scanning ion conductance microscope	SICM	ion current	< 0.1 µm

samples consisting of e-beam lithographed patterns on glass slides, the transmitted signal at each point is measured by a photomultiplier, and an image is constructed. The first experiments found that features < 25 nm could be observed, and the resolution limit of this technique was estimated to be ~ 5–10 nm.

5.2 Photon Scanning Tunneling Microscope (PSTM)

When a beam generated by, for example, an He–Ne laser experiences total internal reflection inside a material (for example, at the face of a prism), an evanescent field exists just outside the planar surface. On a microscopic scale, this evanescent field is modulated by the surface optical properties and structure. If an optical fiber with a very fine end is immersed in the field – at a distance from the surface of the order of the wavelength, λ – internal reflection is locally frustrated in the vicinity of the tip. Light can couple (photon tunneling) to the optical fiber because modification of the internal reflection produces non-zero transmittance. The light then propagates through the optical fiber to a photodetector. This process describes the basis of the photon scanning tunneling microscope (PSTM) (Reddick et al., 1990), which is shown schematically in Fig. 142. With this instrument, a near-field image is obtained by scanning the tip and analyzing the intensity of the propagating light.

For a prism geometry, the transmittance is a function of the dimensionless quantity d/λ, where d is the sample–tip spacing and λ is the wavelength. The transmittance depends on the polarization, the angle of incidence of the beam with respect to the solid–air interface, and the shape of the tip. The decay length, d_I, of the evanescent field away from the surface is of the order of $0.1\,\lambda$–$1.0\,\lambda$, depending on the angle of incidence of the beam and the relative indices of refraction; this scale can be verified by measuring the transmittance as a function of tip–sample separation. From the physical properties and available detectors/electronics, the sensitivity of the instrument provides a vertical resolution of the order of $\lambda/100$.

For imaging, the lateral resolution is ~ $\lambda/10$ if light scattering at the surface can

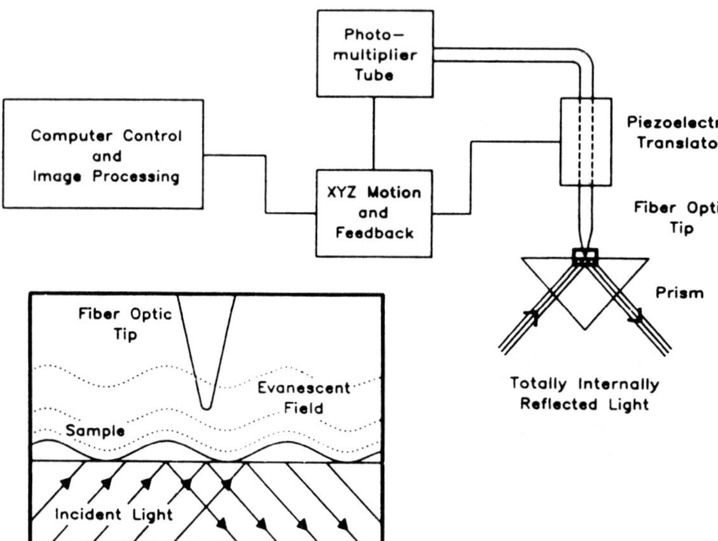

Figure 142. Schematic diagram of PSTM. The total internal reflecting light beam in the prism produces an evanescent field modulated by the sample. Spatial variations in the evanescent field intensity, as probed by a sharp scanning fiber optic probe, produce the image (Reddick et al., 1990).

be eliminated; it is likewise possible to remove interference fringes by using anti-reflecting coatings. One application of the PSTM is for imaging biological molecules which are deposited on the surface of a prism. The presence of the molecule modifies the evanescent field, and these changes are imaged. In Fig. 143, a 5×5 μm image of *E. coli* imbedded in buffer salts is presented showing an apparent height of ~1.5 μm, with visible effects of substrate scattering (Reddick et al., 1990).

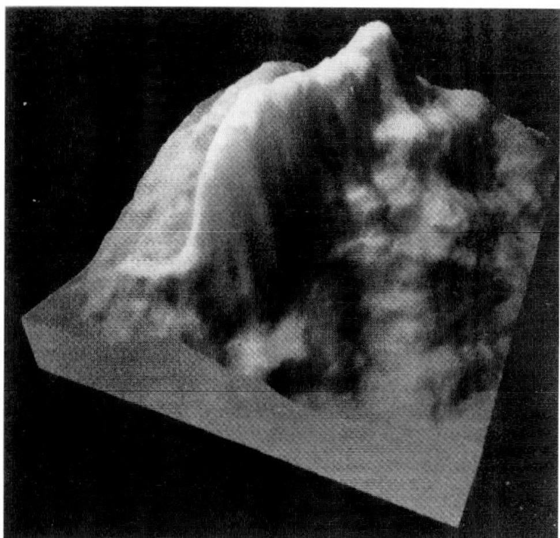

Figure 143. 5 μm × 5 μm PSTM image of *E. coli* embedded in buffer salts. The height range is 3 μm, and the image was obtained with He–Ne laser radiation (Reddick et al., 1990).

The possibility of performing optical spectroscopy – photoluminescence, fluorescence, absorption, Raman – on a local scale is a promising extension of such optical techniques, and time-resolved optical methods can provide a means to study molecular dynamics from a uniquely microscopic point of view.

5.3 Scanning Thermal Profiler (STP)

Two methods of imaging surface temperature using STM techniques have been demonstrated – by using a tip configured as a micro-thermocouple (Williams and Wickramasinghe, 1986), or by directly measuring temperature-dependent differences in the chemical potential between the dissimilar metals of the STM tip and the sample (Williams and Wickramasinghe, 1991), the discussion of which we defer to the next subsection.

The first method produced an instrument called the scanning thermal profiler

(STP), which is based on the fabrication and use of a coaxial micro-thermocouple, as shown in Fig. 144 (Williams and Wickramasinghe, 1986). The microscopic thermocouple junction was formed by coating the shank of the inner conductor with an insulating film, and then by coating the shank and the tip area with a dissimilar metal film so that a miniature thermocouple junction was formed in the tip region. Scanning the tip above a surface with feedback control of the *z*-piezo to maintain constant temperature allows isothermal mapping with sub-micrometer lateral resolution; modulation detection techniques were used to improve the signal-to-noise ratio. An alternative measurement scheme allows topographic mapping based on the variation of thermal conductivities of different materials on the surface. With sequential tip heating/cooling and temperature measurement at each position of the surface, thermal conduction to/from the tip is analyzed. In particular, at a given location in the scan, a current is first passed through the thermocouple to provide Joule/Peltier heating/cooling; second, the temperature of the tip is measured. The topograph of the surface follows that of an isotherm profile; for noise reduction, the gap and sample temperature can be modulated at different frequencies and detected using lock-in techniques. With 10^{-3} °C sensitivity, ~10 Å vertical sensitivity can be achieved. Lateral resolution of < 300 Å was demonstrated in imaging Al strips on a substrate.

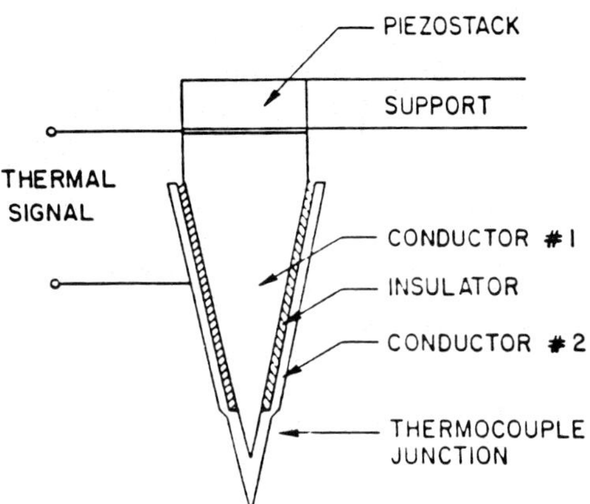

Figure 144. Schematic diagram of an STP micro-thermocouple (Williams and Wickramasinghe, 1986).

5.4 Scanning Chemical Potential Microscope (SCPM)

The scanning chemical potential microscope (SCPM), or "tunneling thermocouple", measures the difference in chemical potential that exists in the region of the tunnel junction between tip and sample made of dissimilar materials and held at different temperatures (Williams and Wickramasinghe, 1990; Williams and Wickramasinghe, 1991). A thermal gradient exists in this region, forming a thermoelectric voltage analogous to the Seeback effect. Instead of having a thermocouple junction formed by the materials in contact, here the junction is the tunnel gap. This is so since the electronic states of the tip and the sample are strongly coupled across the small gap and their electrochemical potentials equalize by electron tunneling in both directions across the gap. Measurement of the potential difference is made in the tunneling range with zero applied bias, and the tip and the sample are held at different temperatures.

The SCPM probes the chemical potential gradient as a result of surfaces consisting of chemically-different materials, or even from atomic scale variations at chemically-homogeneous materials – not variations in the surface temperature on the atomic scale. This arrangement is shown schematically in Fig. 145. The thermoelectric potential for the junction at tem-

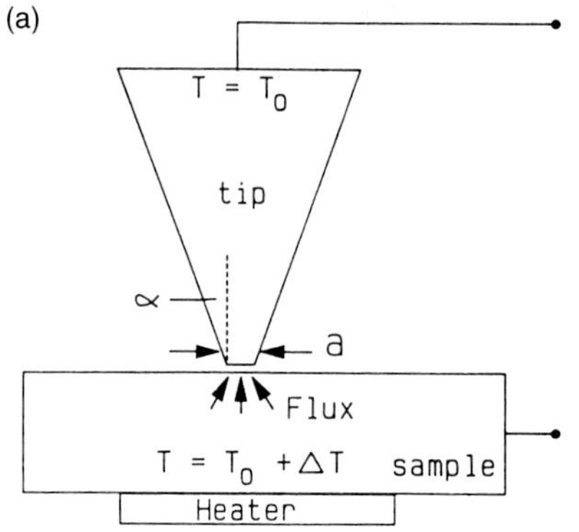

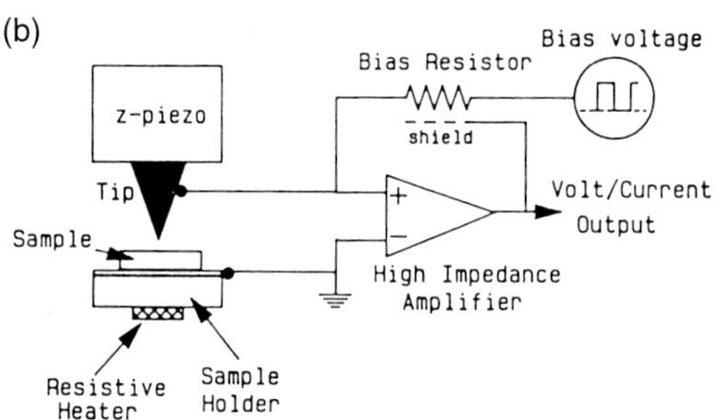

Figure 145. (a) Schematic diagram of SCPM showing thermal characteristics; (b) electronics arrangement to perform sequential STM/SCPM measurements (Williams and Wickramasinghe, 1991).

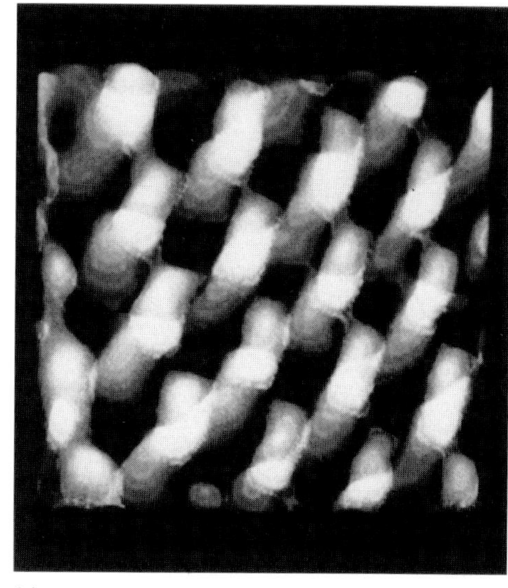
(a)

(b)

Figure 146. SCPM images of (a) graphite – 10 Å × 10 Å – and (b) liquid crystal PYP909 on graphite – 32 Å × 32 Å (Williams and Wickramasinghe, 1990).

perature T and the leads at temperature T_0 is

$$\int_{T_0}^{T} \left(\frac{\partial \mu_{\text{tip}}}{\partial T} - \frac{\partial \mu_{\text{sample}}}{\partial T} \right) \mathrm{d}T \tag{29}$$

where μ_{tip} and μ_{sample} represent respective average chemical potentials. For small temperature differences, $\Delta T_1 = T - T_0$, this integral can be approximated by

$$(\mu'_{\text{tip}}(T)\,\Delta T_1 - \mu'_{\text{sample}}(T))\,\Delta T_1 \tag{30}$$

Assuming that $\partial\mu_{\text{tip}}/\partial T$ is constant, $(\partial\mu_{\text{sample}}/\partial T)\;\delta T$ is the thermoelectric signal, where δT is the temperature differential in the tunnel junction vicinity. Thus the thermoelectric potential is proportional to the chemical potential gradient. Data from the macroscopic Seeback effect give expected values for the thermoelectric potential of $\sim 0.1-1$ mV/K for semiconductors and $\sim 1-10$ μV/K for metals.

Figure 146a shows the thermoelectric image of a (semi-metal) graphite surface where variations of ~ 20 μV are observed. If this image is compared with a simultaneously-acquired STM image, it is found that the "B" sites are peaked in STM, while the "A" sites are peaked in SCPM, illustrating variations in chemical potential over atomic dimensions. Figure 146 b is an image of a layer of liquid crystal deposited on a graphite surface. Here, the coating provides contrast with respect to the graphite since the chemical potential gradient is material-dependent.

5.5. Optical Absorption Microscope (OAM)

The SCPM can also measure local heating that is caused by absorption of optical radiation. Ultimately, imaging surfaces or molecules as a function of wavelength can provide complementary spectroscopic information to bias-dependent electron tunneling measurements, i.e., scanning tunneling spectroscopy (STS). Such spatially-resolved optical absorption measurements using the SCPM to probe local surface temperature variations upon laser irradiation define the optical absorption microscopy (OAM) technique, which is illustrated in Fig. 147 (Weaver et al., 1989). An analogy can be drawn with photothermal deflection spectroscopy (PDS), in which the dissipation of optical excitations at a surface creates thermally-induced stresses which can be seen with optical interferometry. The best attainable resolution for conventional PDS, however, is ~1 μm.

Operation of the scanning contact potential microscope is described above. For OAM measurements, the thermoelectric contrast derived from local heating is due to optical absorption. Light incident from a reference He–Ne laser at 633 nm and a tunable argon-ion laser are separately chopped and detected; the reference beam provides a means for data normalization. Operating in a multiplexed STM/SCPM mode, topographic information and local heating can be correlated. The initial experimental on Au films on mica in air demonstrated two capabilities of the technique – the spatial resolution of OAM images and one type of thermal structure that can be observed. Comparison of STM and OAM images measuring surface expansion and temperature, respectively, at steps set the limit of spatial resolution as ~1 nm. Secondly, the structure observed by STM and OAM from Au on mica and of Al dots on fused-quartz (Fig. 148) may be attributed to subsurface variations due to variations in thermal conduction between the sample and the substrate. Therefore local information on the subsurface can be obtained.

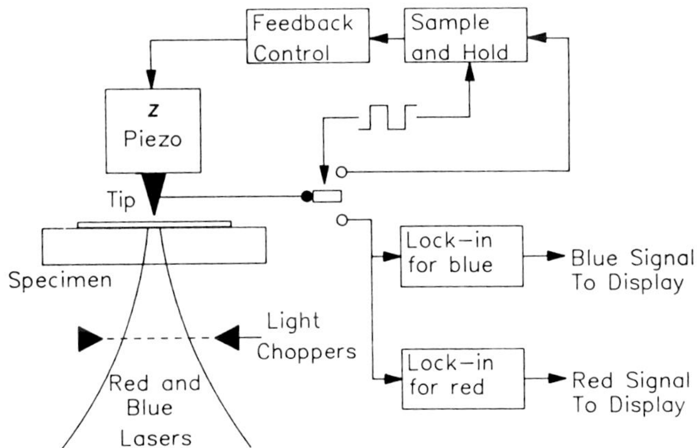

Figure 147. Schematic diagram of OAM. Beams from a 633 nm He–Ne laser and a tunable argon-ion laser are focused on a ~2 μm spot on the sample directly below the tip (Weaver et al., 1989).

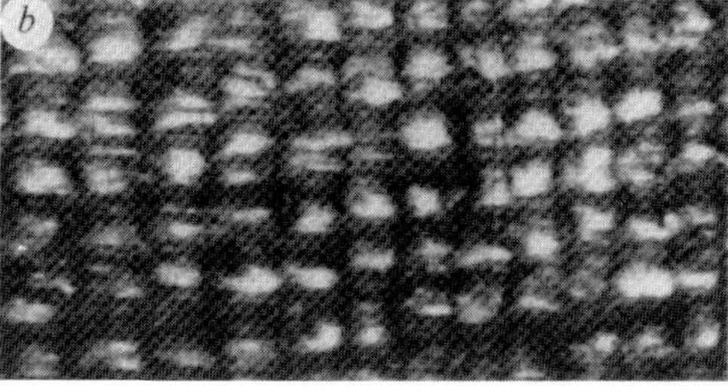

Figure 148. (a) STM image and (b) OAM image at 514 nm of 3000 Å Al dots deposited on a fused-quartz substrate (Weaver et al., 1989).

5.6 Scanning Ion Conductance Microscope (SICM)

Microscopic control and measurement of ion currents near a surface provides a means to probe ion conduction channels in biological systems. The scanning ion conductance microscope (SICM), shown schematically in Fig. 149, has demonstrated < 0.1 μm lateral resolution for imaging the topography and the ion current through pores of model non-conductive membranes immersed in an electrolyte (Hansma et al., 1989). The lateral resolution is essentially governed by the inner diameter of the micropipette which is the core of the instrument. A specially-constructed micropipette filled with an ionic solution is mechanically lowered in close proximity to the sample, which is mounted on a piezoelectric scanner. The ion current between the pipette and the sample is localized by the restricted conductance due to the closeness of the tip to the sample. A topographic scan represents an x-y sweep of the sample while the micropipette is modulated via feedback control of the z-piezo to maintain constant ion current, or an ion current image can be taken at constant height. One of the first demonstrations of the SICM in being used for topographic imaging was for acetate film, where features of the order of 0.2 μm were resolved (Fig. 150).

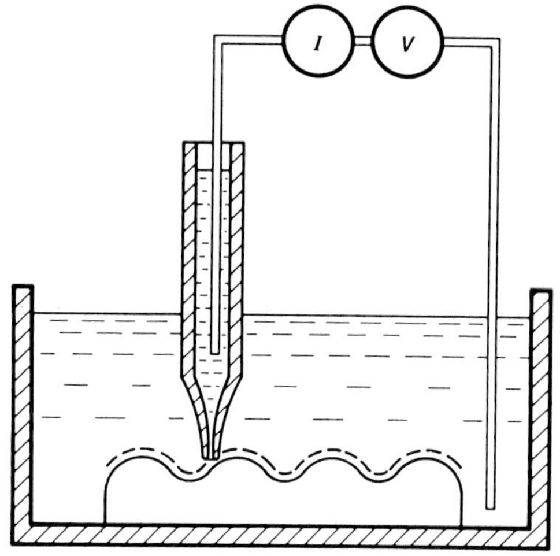

Figure 149. Schematic diagram of SICM (Hansma et al., 1989).

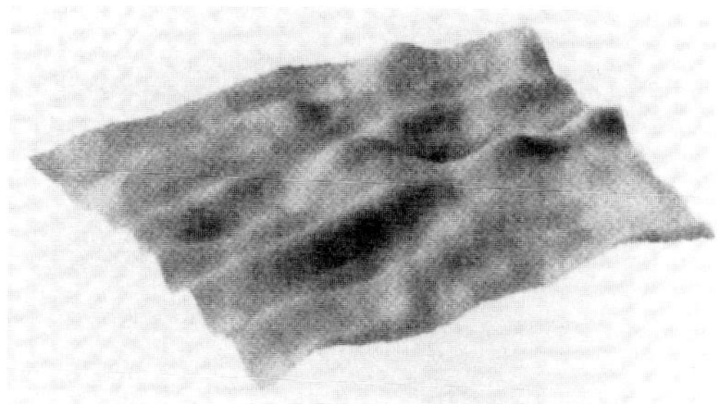

Figure 150. 4 μm × 4 μm SICM image of an acetate film (Hansma et al., 1989).

6 Acknowledgements

I wish to thank Dr. Eric Lifshin and Prof. Robert Cahn for their support and encouragement throughout the writing of the manuscript. I also thank Professor E. W. Plummer, Professor D. Bonnell, Dr. Carl Ventrice, Jr., and Mr. Thomas Mercer for their critical reading of the manuscript. I appreciate the valuable assistance of Mr. Felice Macera in preparing the figures and of Dr. Peter Gregory and Ms. Deborah Hollis at VCH in their editorial and production support.

Financial support has been provided in part by NSF under grant DMR 91-20398 and by the NSF-MRL program at the University of Pennsylvania under grant DMR 91-20668.

7 References

Abraham, D. W., Mamin, H. J., Ganz, E., Clarke, J. (1986), *IBM J. Res. Dev. 30,* 492.

Abraham, D. W., Williams, C. C., Wickramasinghe H. K. (1988), *Appl. Phys. Lett. 53,* 1503–1505.

Akama, Y., Nishimura, E., Sakai, A., Sano, N., Sakurai, T. (1989), *J. de Phys. Colloq., C8,* 211–216.

Akamine, S., Albrecht, T. R., Zdeblick M. J., Quate, C. F. (1989), *IEEE Electron Device Lett. 10,* 490–492.

Akamine, S., Barrett, R. C., Quate, C. F. (1990), *Appl. Phys. Lett. 57,* 316–318.

Albrecht, T. R., Dovek, M. M., Lang, C. A., Grütter, P., Quate, C. F., Kuan, S. W. J., Frank, C. W., Pease, R. F. W. (1988), *J. Appl. Phys. 64,* 1178–1184.

Anders, M., Thaer, M., Lück, M., Heiden, C. (1988), *J. Vac. Sci. Technol. A6,* 436–439.

Anselmetti, D., Weisendanger, R., Guntherodt, H. J. (1989), *Phys. Rev. B39,* 135–138.

Anselmetti, D., Geiser, V., Overney, G., Wiesendanger, R., Güntherodt, H. (1990), *Phys. Rev. B42,* 1848–1851.

Avouris, P., Lyo, I. (1992), *Appl. Surf. Sci. 60/61,* 426–436.

Avouris, P., Wolkow, R. (1989a), *Phys. Rev. B39,* 5091.

Avouris, P., Wolkow, R. (1989b), *Appl. Phys. Lett. 55,* 1074.

Avouris, P., Lyo, I.-W., Bozso, F., Kaxiras, E. (1990), *J. Val. Sci. Technol. A8,* 3405–3411.

Avouris, P., Lyo, I.-W., Bozso, F. (1991), *J. Vac. Sci. Technol. B9,* 424–430.

Baratoff, A., Persson, B. N. J. (1988), *J. Vac. Sci. Technol. A6,* 331–335.

Bardeen, J. (1961), *Phys. Rev. Lett. 6,* 57–59.

Baró, A. M., Miranda, R., Alamán, J., García, N., Binnig, G., Rohrer, H., Gerber, C., Carrascosa, J. L. (1985), *Nature 315,* 253–254.

Barrett, R. C., Quate, C. F. (1990), *J. Vac. Sci. Technol. A8,* 400–402.

Baski, A. A., Nogami, J., Quate, C. F. (1990), *J. Vac. Sci. Technol. A8,* 245–248.

Baski, A. A., Nogami, J., Quate, C. F. (1991), *J. Vac. Sci. Technol. A9,* 1946–1950.

Batra, I. P. Ciraci, S. (1988), *J. Vac. Sci. Technol. A6,* 313–318.

Batra, I. P. García, N., Rohrer, H., Salemink, H., Stoll, E., Ciraci, S. (1987), *Surf. Sci. 181,* 126–138.

Becker, R. S., Golovchenko, J. A., Swartzentruber, B. S. (1985a), *Phys. Rev. Lett. 55,* 987–990.

Becker, R. S., Golovchenko, J. A., Haman, D. R., Swartzentruber, B. S. (1985b), *Phys. Rev. Lett. 55,* 2032–2034.

Becker, R. S., Golovchenko, J. A., Higashi, G. S., Swartzentruber, B. S. (1986), *Phys. Rev. Lett. 57,* 1020–1023.

Becker, R. S., Golovchenko, J. A., Swartzentruber, B. S. (1987), *Nature 325,* 419–421.

Becker, R. S., Swartzentruber, B. S., Vickers, J. S., Klitsner, T. (1989), *Phys. Rev. B39,* 1633–1647.

Becker, R. S., Higashi, G. S., Chabal, Y. J., Becker, A. J. (1990), *Phys. Rev. Lett. 65,* 1917–1920.

Bednorz, J. G., Müller, K. A. (1986), *Z. Phys. B64,* 189.

Bedrossian, P., Meade, R. D., Mortensen, K., Chen, D. M., Golovchenko, J. A., Vanderbilt, D. (1989), *Phys. Rev. Lett. 63,* 1257–1260.

Beebe, Jr., T. P., Wilson, T. E., Ogletree, D. F., Katz, J. E., Balhorn, R., Salmeron, M. B., Siekhaus, W. J. (1989), *Science 243,* 370–372.

Behm, R. J., Hösler, W., Ritter, E., Binnig, G. (1986), *Phys. Rev. Lett. 56,* 228–231.

Bell, A. E., Rao, K., Swanson, L. W. (1988), *J. Vac. Sci. Technol. B6,* 306–310.

Bell, L. D., Kaiser, W. J. (1988), *Phys. Rev. Lett. 61,* 2368–2371.

Bell, L. D., Hecht, M. H., Kaiser, W. J., Davis, L. C. (1990), *Phys. Rev. Lett. 64,* 2679–2682.

Bendixen, C., Besenbacher, F., Lægsgaard, E., Stensgaard, I., Thomsen, B., Westergard, O. (1990), *J. Vac. Sci. Technol. A8,* 703–705.

Berndt, R., Gimzewski, J. K., Johansson, P. (1991), *Phys. Rev. Lett. 67,* 3796–3799.

Berthe, R., Halbritter, J. (1991), *Phys. Rev. B43,* 6880–6884.

Besocke, K. (1987), *Surf. Sci. 181,* 145–153.

Binnig, G. (1985), *Surf. Sci. Lett. 157,* L373.

Binnig, G., Rohrer, H. (1983), *Surf. Sci. 126,* 236–244.

Binnig, G., Rohrer, H. (1986), *IBM J. Res. Dev. 30,* 355–369.

Binnig, G., Rohrer, H. (1987), Rev. Mod. Phys. 59, 615–625.

Binnig, G., Smith, D. P. E. (1986), *Rev. Sci. Instrum. 57,* 1688–1689.

Binnig, G., Rohrer, H., Gerber, C., Weibel, H. (1982a), *Phys. Rev. Lett. 49,* 57–61.

Binnig, G., Rohrer, H., Gerber, C., Weibel, H. (1982b), *Appl. Phys. Lett. 40,* 178–180.

Binnig, G., Rohrer, H., Gerber, C., Weibel, H. (1983a), *Surf. Sci. Lett. 131,* L379–L384.

Binnig, G., Rohrer, H., Gerber, C., Weibel, H. (1983b), *Phys. Rev. Lett. 50,* 120–123.

Binnig, G., García, N., Rohrer, H., Soler, J. M., Flores, F. (1984), *Phys. Rev. B30,* 4816–4818.

Binnig, G., Frank, K. H., Fuchs, H., García, N., Reihl, B., Rohrer, H., Salvan, F., Williams, A. R. (1985), *Phys. Rev. Lett. 55,* 991–994.

Binnig, G., Quate, C. F., Gerber, C. (1986), *Phys. Rev. Lett. 56,* 930–933.

Blackford, B. L., Dahn, D. C., Jericho, M. H. (1987), *Rev. Sci. Instrum. 58,* 1343–1348.

Boland, J. J. (1990), *Phys. Rev. Lett. 65,* 3325–3328.

Boland, J. J. (1991a), *J. Vac. Sci. Technol. B9,* 764–769.

Boland, J. J. (1991b), *Phys. Rev. Lett. 67,* 1539–1542.

Bonnell, D. A. (1988), *Mater. Sci. Eng. A105/106,* 55–63.

Bonnell, D. A., Clarke, D. R. (1988), *J. Am. Ceram. Soc. 71,* 629–637.

Bourgiugnon, B., Carleton, K. L., Leone, S. R. (1988a), *Surf. Sci. 204,* 455–472.

Bourgiugnon, B., Smilgys, R. V., Leone, S. R. (1988b), *Surf. Sci. 204,* 473–484.

Bringans, R. D., Uhrberg, R. I. G., Bachrach, R. Z., Northrup, J. E. (1985), *Phys. Rev. Lett. 55,* 533–536.

Brocks, G., Kelly, P. J., Car, R. *Phys. Rev. Lett. 70,* 2786–2788.

Brommer, K. D., Needels, M., Larson, B. E., Joannopoulos, J. D. (1992), *Phys. Rev. Lett. 68,* 1355–1358.

Burk, B., Thomson, T. E., Zettl, A., Clarke, J. (1991), *Phys. Rev. Lett. 66,* 3040–3043.

Burnham, N. A., Colton, R. J. (1989), *J. Vac. Sci. Technol. A7,* 2906–2913.

Burnham, N. A., Dominguez, D. D., Mowrey, R. L. (1990), *Phys. Rev. Lett. 64,* 1931–1934.

Carlsson, P., Holmström, B., Kita, H., Uosaki, K. (1990), *Surf. Sci. 237,* 280–290.

Chadi, D. J. (1979), *Phys. Rev. Lett. 43,* 43–47.

Chen, C. J. (1992), *Appl. Phys. Lett. 60,* 132–134.

Chua, F. M., Kuk, Y., Silverman, P. J. (1989), *Phys. Rev. Lett. 63,* 386–389.

Ciraci, S., Tekman, E. (1989), *Phys. Rev. B40,* 11969–11972.

Ciraci, S., Baratoff, A., Batra, I. P. (1990), *Phys. Rev. B42,* 7618–7621.

Coleman, R. V., Drake, B., Hansma, P. K., Slough, G. (1985), *Phys. Rev. Lett. 55,* 394–397.

Coleman, R. V., Giambattista, B., Hansma, P. K., Johnson, A., McNairy, W. W., Slough, C. G. (1988), *Adv. Phys.,* 599–644.

Coleman, R. V., Dai, Z., McNairy, W. W., Slough, C. G., Wang, C. (1992), *Appl. Surf. Sci. 60/61,* 485–490.

Colton, R. J., Baker, S. M., Drisoll, R. J., Youngquist, M. G., Baldeschwieler, J. B., Kaiser, W. J. (1987), *J. Vac. Sci. Technol. A6,* 349–353.

Coombs, J. H., Pethica, J. B., Welland, M. E. (1988), *Thin Solid Films 159,* 293–299.

Coulman, D. J., Wintterlin, J., Behm, R. J., Ertl, G. (1990), *Phys. Rev. Lett. 64,* 1761–1764.

Crandall, B. C., Lewis, J. (Eds.) (1992), *Nanotechnology: Research and Perspectives.* Cambridge, MA: MIT Press.

Cricenti, A., Selci, S., Chiarotti, G., Amaldi, F. (1991), *J. Vac. Sci. Technol. B2,* 1285–1287.

Dagata, J. A., Schneir, J., Harary, H. H., Evans, C. J., Postek, M. T., Bennett, J. (1990), *Appl. Phys. Lett. 56,* 2001–2003.

Dahn, D. D., Cake, K., Hale, L. R. (1992), *Ultramicroscopy 42–44,* 1222–1227.

Demuth, J. E., Hamers, R. J., Tromp, R. M., Welland, M. E. (1986), *J. Vac. Sci. Technol. 4,* 1320–1323.

Denley, D. R. (1990), *Ultramicroscopy 33,* 83–92.

Dietz, P., Herrmann, K. (1990), *Surf. Sci. 232,* 339–345.

DiNardo, N. J., Demuth, J. E., Thompson, W. A., Avouris, P. (1985), *Phys. Rev. B31,* 4077.

DiNardo, N. J., Maeda Wong, T., Plummer, E. W. (1990), *Phys. Rev. Lett. 65,* 2177–2180.

Ding, Y. G., Chan, C. T., Ho, K. M. (1991), *Phys. Rev. Lett. 67,* 1454–1457.

Dittrich, R., Heiden, C. (1988), *J. Vac. Sci. Technol. A6,* 263–265.

Dovek, M. M., Lang, C. A., Nogami, J., Quate, C. F. (1989), *Phys. Rev. B40,* 11973–11975.

Dragoset, R. A., First, P. N., Stroscio, J. A., Pierce, D. T., Celotta, R. J. (1989), *Mater. Res. Soc. Symp. Proc. 151,* 193–198.

Drechsler, M., Blackford, B. L., Putnam, A. M., Jericho, M. H. (1989), *J. de Phys. Colloq. C8,* 223–228.

Eigler, D. M., Schweizer, E. K. (1990), *Nature 344,* 524–526.

Eigler, D. M., Lutz, C. P., Rudge, W. E. (1991a), *Nature 352,* 600–603.

Eigler, D. M., Weiss, P. S., Schweizer, E. K., Lang, N. D. (1991b), *Phys. Rev. Lett. 66,* 1189–1192.

Elings, V., Edstrom, R. D., Meinke, M. H., Yang, X., Yang, R., Evans, D. F. (1990), *J. Vac. Sci. Technol. A8,* 652–654.

Elrod, S. A., de Lozanne, A. L., Quate, C. F. (1984), Preprint.

Emch, R., Clivax, X., Taylor-Denes, C., Vaudaus, P., Descouts, P. (1990), *J. Vac. Sci. Technol. A8,* 655–658.

Erlandson, R., McClelland, G. M., Mate, C. M., Chiang, S. (1988), *J. Vac. Sci. Technol. A6,* 266–270.

Everson, M. P., Davis, L. C., Jaklevic, R. C., Shen, W. (1991), *J. Vac. Sci. Technol. B9,* 891–896.

Fan, F.-R. F., Bard, A. J. (1990), *J. Phys. Chem. 94,* 3761–3766.

Fauster, T., Himpsel, F. J. (1983), *J. Vac. Sci. Technol. A1,* 1111–1114.

Feenstra, R. M. (1989), *Phys. Rev. Lett. 63,* 1412–1415.

Feenstra, R. M., Lutz, M. A. (1990), *Phys. Rev. B42,* 5391–5394.

Feenstra, R. M., Lutz, M. A. (1991 a), *Surf. Sci. 243,* 151–165.

Feenstra, R. M., Lutz, M. A. (1991 b), *J. Vac. Sci. Technol. B9,* 716.

Feenstra, R. M., Mårtensson, P. (1988), *Phys. Rev. Lett. 61,* 447–450.

Feenstra, R. M., Thompson, W. A., Fein, A. P. (1986), *Phys. Rev. Lett. 56,* 608–611.

Feenstra, R. M., Stroscio, J. A., Fein, A. P. (1987 a), *Surf. Sci. 181,* 292–306.

Feenstra, R. M., Stroscio, J. A., Tersoff, J., Fein, A. P. (1987 b), *Phys. Rev. Lett. 58,* 1192–1195.

Feenstra, R. M., Mårtensson, P., Ludeke, R. (1989 a), *Mater. Res. Soc. Symp. Proc. 138,* 305–314.

Feenstra, R. M., Mårtensson, P., Stroscio, J. A. (1989 b), in: Metallization and Metal-Semiconductor Interfaces. New York: Plenum.

Feenstra, R. M., Slavin, A. J., Held, G. A., Lutz, M. A. (1991), *Phys. Rev. Lett. 66,* 3257–3260.

Fein, A. P., Kirtley, J. R., Feenstra, R. M. (1987), *Rev. Sci. Instrum. 58,* 1806–1810.

Feltz, A., Memmert, U., Behm, R. J. (1992), *Chem. Phys. Lett. 192,* 271–276.

Fernandez, A., Hallen, H. D., Huang, T., Buhrman, R. A., Silcox, J. (1990), *Appl. Phys. Lett. 57,* 2826–2828.

First, P. N., Dragoset, R. A., Stroscio, J. A., Celotta, R. J., Feenstra, R. M. (1989 a), *J. Vac. Sci. Technol. A7,* 2868–2872.

First, P. N., Stroscio, J. A., Dragoset, R. A., Pierce, D. T., Celotta, R. J. (1989 b), *Phys. Rev. Lett. 63,* 1416–1419.

Foster, J. S., Frommer, J. E. (1988), *Nature 333,* 542–545.

Foster, J. S., Frommer, J. E., Arnett, P. C. (1988), *Nature 331,* 324–326.

Frohn, J., Wolf, J. F., Besocke, K., Teske, M. (1989), *Rev. Sci. Instrum. 60,* 1200–1201.

Frohn, J., Giesen, M., Poensgen, M., Wolf, J. F., Ibach, H. (1991), *Phys. Rev. Lett. 67,* 3543–3546.

Fuchs, H., Akari, S., Dransfeld, K. (1990), *Z. Phys. B – Condensed Matter 80,* 389–392.

Fujiwara, I., Ishimoto, C., Seto, J. (1991), *J. Vac. Sci. Technol. B9,* 1148–1153.

Gammie, G., Hubacek, J. S., Skala, S. L., Brockenbrough, R. T., Tucker, J. R., Lyding, J. W. (1989 a), *Phys. Rev. B40,* 9529–9532.

Gammie, G., Hubacek, J. S., Skala, S. L., Brockenbrough, R. T., Tucker, J. R., Lyding, J. W. (1989 b), *Phys. Rev. B40,* 11965–11968.

Gammie, G., Hubacek, J. S., Skala, S. L., Tucker, J. R., Lyding, J. W. (1991), *J. Vac. Sci. Technol. B9,* 1027–1031.

Garnaes, J., Schwartz, D. K., Viswanathan, R., Zasadizinski, J. A. N. (1992), *Nature 357,* 54–57.

Gehrtz, M., Strecker, H., Grimm, H. (1988), *J. Vac. Sci. Technol. A6,* 432–435.

Gerber, C., Binnig, G., Fuchs, H., Marti, O., Rohrer, H. (1986), *Rev. Sci. Instrum. 57,* 221–224.

Giaver, I. (1960), *Phys. Rev. Lett. 5,* 147–148.

Gilbert, S. E., Kennedy, J. H. (1990), *Surf. Sci. Lett. 225,* L1–L7.

Gimzewski, J. K., Möller, R., Pohl, D. W., Schittler, R. R. (1987), *Surf. Sci. 189/190,* 15–23.

Göddenhenrich, T., Lemke, H., Hartmann, U., Heiden, C. (1990 a), *J. Vac. Sci. Technol. A8,* 383–387.

Göddenhenrich, T., Lemke, H., Hartmann, U., Heiden, C. (1990 b), *Appl. Phys. Lett. 56,* 2578–2580.

Gómez, J., Vázquez, L., Baró, A. M., García, N., Pedriel, C. L., Triaca, W. E., Arvia, A. J. (1986), *Nature 323,* 612–616.

Gould, S. A. C., Drake, B., Prater, C. B., Weisenhorn, A. L., Manne, S., Hansma, H. G., Hansma, P. K., Massie, J., Longmire, M., Elings, V., Dixon Northern, B., Mukergee, B., Peterson, C. M., Stoeckenius, W., Albrecht, T. R., Quate, C. F. (1990), *J. Vac. Sci. Technol. A8,* 369–373.

Gritsch, T., Coulman, D., Behm, R. J., Ertl, G. (1989), *Phys. Rev. Lett. 63,* 1086–1089.

Grunze, M., Unertl, W. N., Gnanarajan, S., French, J. (1988), *Mater. Res. Soc. Symp. Proc. 108,* 189–199.

Gryko, J., Allen, R. E. (1992), *Ultramicroscopy 42–44,* 793–800.

Guckenberger, R., Wiegräbe, W., Hillebrand, A., Hartmann, T., Wang, X., Baumeister, W. (1989), *Ultramicroscopy 31,* 327–332.

Guckenberger, R., Hacker, B., Hartmann, T., Scheybani, T., Wang, Z., Wiegräbe, W., Baumeister, W. (1991), *J. Vac. Sci. Technol. B2,* 1227–1230.

Gygi, F., Schluter, M. (1990), *Phys. Rev. Lett. 41,* 822–825.

Habib, K., Eling, V., Wu, C. (1990), *J. Mater. Sci. Lett. 9,* 1194.

Hachiya, T., Itaya, K. (1992), *Ultramicroscopy 42–44,* 445–452.

Hallmark, V. M., Chiang, S., Rabolt, J. F., Swalen, J. D., Wilson, R. J. (1987 a), *Phys. Rev. Lett. 59,* 2879–2882.

Hallmark, V. M., Leone, A., Chiang, S., Swalen, J. D., Rabolt, J. F. (1987 b), *Polym. Prepr. (Am. Chem. Soc., Div. Polym. Chem.) 28,* 22–23.

Hameroff, S., Simic-Krstic, Y., Vernetti, L., Lee, Y. C., Sarid, D., Wiedmann, J., Elings, V., Kjoller,

K., McCuskey, R. (1990), *J. Vac. Sci. Technol. A8,* 687–691.

Hamers, R. J. (1989), Atomic-resolution surface spectroscopy with the scanning tunneling microscope. *Annu. Rev. Phys. Chem.* 531–559.

Hamers, R. J., Cahill, D. G. (1990), *Appl. Phys. Lett. 57,* 2031–2033.

Hamers, R. J., Demuth, J. E. (1988), *Phys. Rev. Lett. 60,* 2527–2530.

Hamers, R. J., Markert, K. (1990), *Phys. Rev. Lett. 64,* 1051–1054.

Hamers, R. J., Tromp, R. M., Demuth, J. E. (1986a), IBM Res. Rep.

Hamers, R. J., Tromp, R. M., Demuth, J. E. (1986b), *Phys. Rev. B34,* 5343–5357.

Hamers, R. J., Tromp, R. M., Demuth, J. E. (1986c), *Phys. Rev. Lett. 56,* 1972–1975.

Hamers, R. J., Avouris, P., Bozso, F. (1987), *Phys. Rev. Lett. 59,* 2071–2074.

Hansma, H. G., Weisenhorn, A. L., Gould, S. A. C., Sinsheimer, R. L., Gaub, H. E., Stucky, G. D., Zaremba, C. M., Hansma, P. K. (1991), *J. Vac. Sci. Technol. B9,* 1282–1284.

Hansma, P. K., Tersoff, J. (1987), *J. Appl. Phys. 61,* R1–R23.

Hansma, P. K., Drake, B., Marti, O., Goud, S. A. C., Prater, C. B. (1989), *Science 243,* 641–643.

Hawley, M. E., Gray, K. E., Terris, B. D., Wang, H. H., Carlson, K. D., Williams, J. M. (1986), *Phys. Rev. Lett. 57,* 629–632.

Headrick, R. L., Robinson, I. K., Vlieg, E., Feldman, L. C. (1989), *Phys. Rev. Lett. 63,* 1253–1256.

Hecht, M. H., Bell, L. D., Kaiser, W. J., Davis, L. C. (1990), *Phys. Rev. B42,* 7663–7666.

Heslinga, D. R., Weitering, H. H., van der Werf, D. P., Klapwijk, T. M., Hibma, T. (1990), *Phys. Rev. Lett. 64,* 1589–1592.

Hess, H. F., Robinson, R. B., Dynes, R. C., Valles Jr., J. M., Waszczak, J. V. (1989), *Phys. Rev. Lett. 62,* 214–216.

Hess, H. F., Robinson, R. B., Waszczak, J. V. (1990), *Phys. Rev. Lett. 64,* 2711–2714.

Hibino, H., Fukuda, T., Suzuki, M., Homma, Y., Sato, T., Iwatsuki, M., Miki, K., Tokumoto, H. (1993), *Phys. Rev. B47,* 13027–13030.

Higashi, G. S., Chabal, Y. J., Trucks, G. W., Ragahavachari, K. (1990), *Phys. Rev. Lett. 56,* 656–658.

Himpsel, F. J., Fauster, T. (1984), *J. Vac. Sci. Technol. A2,* 815–821.

Hipps, K. W., Fried, G., Fried, D. (1990), *Rev. Sci. Instrum. 61,* 1869–1873.

Höfer, U., Morgen, P., Wurth, W., Umbach, E. (1989), *Phys. Rev. B40,* 1130–1144.

Hörber, J. K. H., Schuler, F. M., Witzemann, V., Schröter, K. H., Müller, H., Ruppersberg, J. P. (1991), *J. Vac. Sci. Technol. B9,* 1214–1218.

Huang, Z., Allen, R. E. (1992), *Ultramicroscopy 42–44,* 97–104.

Ibach, H., Mills, D. L. (1982): *Electron Energy Loss Spectroscopy and Surface Vibrations.* New York: Academic Press.

Ide, T., Nishimori, T., Ichinokawa, T. (1989), *Surf. Sci. 209,* 333–344.

Ishiki, N., Kobayashi, K., Tsukada, M. (1990), *Surf. Sci. Lett. 238,* L439–L445.

Israelachvili, J. N., Adams, G. E. (1978), *J. Chem. Soc., Faraday Trans. 1,* 74, 975.

Itaya, K., Tomita, E. (1989), *Chem. Lett. 1989,* 285.

Iwatsuki, M., Kitamura, S.-i. (1990), *JEOL News 28E,* 24–28.

Iwatsuki, M., Kitamura, S., Sato, T., Sueyoshi, T. (1992), *Appl. Surf. Sci. 60/61,* 580–586.

Jaklevic, R. C., Elie, L. (1988), *Phys. Rev. Lett. 60,* 120–123.

Jensen, F., Besenbacher, F., Lægsgaard, E., Stensgaard, I. (1990), *Phys. Rev. B42,* 9206–9209.

Jericho, M. H., Blackford, B. L., Dahn, D. C., Frame, C., Maclean, D. (1990), *J. Vac. Sci. Technol. A8,* 661–666.

Jiang, M. H., Ventrice Jr., C. A., Scoles, K. J., Tyagi, S., DiNardo, N. J., Rothwarf, A. (1991), *Physica C183,* 39–50.

Kaiser, W. J., Bell, L. D. (1988), *Phys. Rev. Lett. 60,* 1406–1409.

Kaiser, W. J., Jaklevic, R. C. (1987), *Surf. Sci. 182,* L227–L233.

Kaiser, W. J., Bell, L. D., Hecht, M. H., Grunthaner, F. J. (1989), *J. Vac. Sci. Technol. B7,* 945–949.

Keller, D. J., Chih-Chung, C. (1992), *Surf. Sci. 268,* 333–339.

Kelty, S. P., Lieber, C. M. (1989), *Phys. Rev. B40,* 5856–5859.

Kirk, M. D., Nogami, J., Baski, A. A., Mitzi, D. B., Kapitulnik, A., Geballe, T. H., Quate, C. F. (1988), *Science 242,* 1673–1675.

Kirtley, J. R., Feenstra, R. M., Fein, A. P., Raider, S. I., Gallagher, W. J., Sandstrom, R., Dinger, T., Shafer, M. W., Koch, R., Laibowitz, R., Bumble, B. (1988a), *J. Vac. Sci. Technol. A6,* 259–262.

Kirtley, J. R., Washburn, S., Brady, M. J. (1988b), *Phys. Rev. Lett. 60,* 1546–1549.

Kitamura, S.-i., Sato, T., Iwatsuki, M. (1991), *Nature 351,* 215–217.

Knall, J., Sundgren, J., Hansson, G. V., Greene, J. E. (1986), *Surf. Sci. 16,* 512–538.

Kubby, J. A., Greene, W. J. (1992), *Phys. Rev. Lett. 68,* 329–332.

Kuk, Y., Silverman, P. J. (1986), *Appl. Phys. Lett. 48,* 1597–1599.

Kuk, Y., Silverman, P. J. (1989), *Rev. Sci. Instrum. 60,* 165–180.

Land, T. A., Michely, T., Behm, R. J., Hemminger, J. C., Comsa, G. (1991), *Appl. Phys. A53,* 414–417.

Lang, N. D. (1985), *Phys. Rev. Lett. 55,* 230–233.

Lang, N. D. (1986), *Phys. Rev. B34,* 5947–5950.

Lang, N. D. (1987a), *Phys. Rev. Lett. 58,* 45–48.

Lang, N. D. (1987b), *Phys. Rev. B36,* 8173–8176.

Lang, N. D., Kohn, W. (1970), *Phys. Rev. B1,* 4555–4568.

Lieber, C. M., Kim, Y. (1991), *Thin Solid Films 206,* 355–359.

Lin, C. W., Fan, F. R., Bard, A. J. (1987), *J. Electrochem. Soc. 134,* 1038.

Lin, J. N., Drake, B., Lea, A. S., Hansma, P. K., Andrade, J. D. (1990), *Langmuir 6,* 509–511.

Lindsay, S. M., Barris, B. (1988), *J. Vac. Sci. Technol. A6,* 544–547.

Lindsay, S. M., Thundat, T., Nagahara, L., Knipping, U., Rill, R. L. (1989), *Science 244,* 1063–1064.

Liu, J.-X., Wan, J.-C., Goldman, A. M., Chang, Y. C., Jiang, P. Z. (1991), *Phys. Rev. Lett. 67,* 2195–2198.

Ludeke, R. (1993), *Phys. Rev. Lett. 70,* 214–217.

Ludeke, R., Prietsch, M., Samsavar, A. (1991), *J. Vac. Sci. Technol. B9,* 2342–2348.

Lyding, J. W., Skala, S., Hubacek, J. S., Brockenbrough, R., Gammie, G. (1988), *Rev. Sci. Instrum. 59,* 1897–1902.

Lyo, I.-W., Avouris, P. (1989), *Science 245,* 1369–1371.

Lyo, I.-W., Avouris, P. (1991), *Science 253,* 173–176.

Lyo, I.-W., Kaxiras, E., Avouris, P. (1989), *Phys. Rev. Lett. 63,* 1261–1264.

Mainsbridge, B., Thundat, T. (1991), *J. Vac. Sci. Technol. B9,* 1259–1262.

Mamin, H. J., Ganz, E., Abraham, D. W., Thomson, R. E., Clarke, J. (1986), *Phys. Rev. B34,* 9015–9018.

Mamin, H. J., Rugar, D., Stern, J. E., Fontana Jr., R. E., Kasiraj, P. (1989), *Appl. Phys. Lett. 55,* 318–320.

Mamin, H. J., Guenther, P. H., Rugar, D. (1990), *Phys. Rev. Lett. 65,* 2418–2421.

Manne, S., Butt, H. J., Gould, S. A. C., Hansma, P. K. (1990), *Appl. Phys. Lett. 56,* 1758–1759.

Marti, O., Drake, B., Hansma, P. K. (1987), *Appl. Phys. Lett. 51,* 484–486.

Martin, Y., Wickramasinghe, H. K. (1987), *Appl. Phys. Lett. 50,* 1455–1457.

Martin, Y., Williams, C. C., Wickramasinghe, H. K. (1987), *J. Appl. Phys. 61,* 4723–4729.

Masai, J., Shibata, T., Kagawa, Y., Kondo, S. (1992), *Ultramicroscopy 42–44,* 1194–1199.

Mate, C. M., McClelland, G. M., Erlandsson, R., Chiang, S. (1987), *Phys. Rev. Lett. 59,* 1942–1945.

Matsumoto, T., Tanaka, H., Kawai, T., Kawai, S. (1992), *Surf. Sci. Lett. 278,* L153–158.

McCord, M. A., Pease, R. F. W. (1987), *J. Vac. Sci. Technol. B5,* 430–433.

McMaster, T. J., Carr, H., Miles, M. J., Cairns, P., Morris, V. J. (1990), *J. Vac. Sci. Technol. A8,* 672–674.

Mercer, T. W., Pizzillo, T. J., Ventrice Jr., C. A., DiNardo, N. J. (1991), unpublished results.

Meyer, E., Heinzelmann, H., Brodbeck, D., Overney, G., Overney, R., Howald, L., Hug, H., Jung, T., Hidber, H., Güntherodt, H. (1991), *J. Vac. Sci. Technol. B9,* 1329–1332.

Miki, K., Morita, Y., Tokumoto, H., Sato, T., Iwatsuki, M., Suzuki, M., Fukuda, T. (1992), *Ultramicroscopy 42–44,* 851–857.

Miles, M., J., McMaster, T., Carr, H. J., Tatham, A. S., Shewry, P. R., Field, J. M., Belton, P. S., Jeenes, D., Hanley, B., Whittam, M., Cairns, P., Morris, V. J., Lambert, N. (1990), *J. Vac. Sci. Technol. A8,* 698–702.

Mizes, H. A., Park, S.-I., Harrison, W. A. (1987), *Phys. Rev. B36,* 4491–4494.

Mo, Y. (1991), *Surf. Sci. 248,* 313–320.

Mo, Y. W. (1991), *Phys. Rev. Lett. 66,* 1998–2001.

Mo, Y., Swartzentruber, B. S., Kariotis, P., Webb, M. B., Lagally, M. G. (1989), *Phys. Rev. Lett. 63,* 2393–2396.

Mo, Y., Savage, D. E., Swartzentruber, B. S., Lagally, M. G. (1990a), *Phys. Rev. Lett. 65,* 1020–1023.

Mo, Y. W., Kariotis, R., Swartzentruber, B. S., Webb, M. B., Lagally, M. G. (1990b), *J. Vac. Sci. Technol. B8,* 232–236.

Moreland, J., Ekin, J. W., Goodrich, L. F., Capobianco, T. E., Clark, A. F. (1987), *Phys. Rev. B35,* 8856–8857.

Morgan, B. A., Stupian, G. W. (1991), *Rev. Sci. Instrum. 62,* 3112–3113.

Muralt, P., Meier, H., Pohl, D. W., Salemink, H. W. M. (1987), *Appl. Phys. Lett. 50,* 1352–1354.

Nakanishi, K., Shiba, H. (1977), *J. Phys. Soc. Jpn. 43,* 1893.

Neubauer, G., Cohen, S. R., McClelland, G. M., Horne, D., Mate, C. M. (1990), *Rev. Sci. Instrum. 61,* 2296–2308.

Niehus, H., Raunau, W., Besocke, K., Spitzl, R., Comsa, G. (1990), *Surf. Sci. Lett. 225,* L8–L14.

Nogami, J., Baski, A. A., Quate, C. F. (1991), *Phys. Rev. B44,* 1415–1418.

Nomura, K., Ichimura, K. (1990), *J. Vac. Sci. Technol. A8,* 504–507.

Northrup, J. E., Schabel, M. C., Karlsson, C. J., Uhrberg, R. I. G. (1991), *Phys. Rev. B44,* 13799–13802.

Ohtani, H., Wilson, R. J., Chiang, S., Mate, C. M. (1988), *Phys. Rev. Lett. 60,* 2398–2401.

Okano, M., Kajimura, K., Wakiyama, S., Sakai, F., Mizutani, W., Ono, M. (1987), *J. Vac. Sci. Technol. A5,* 3313–3320.

Okayama, S., Komuro, M., Mizutani, W., Tokumoto, H., Okano, M., Shimzu, K., Kobayashi, Y., Matsumoto, F., Wakiyama, S., Shigeno, M., Sakai, F., Fujiwara, S., Kitamura, O., Ono, M., Kajimura, K. (1988), *J. Vac. Sci. Technol. A6,* 440–445.

Olk, C. H., Heremans, J., Dresselhaus, M. S., Speck, J. S., Nicholls, J. T. (1990), *Phys. Rev. B42,* 7524–7529.

Olmstead, M. A., Bringans, R. D., Uhrberg, R. I. G., Bachrach, R. Z. (1986), *Phys. Rev. B34,* 6041–6044.

Oppenheim, I. C., Trevor, D. J., Chidsey, C. E. D., Trevor, P. L., Sieradzki, K. (1991), *Science 254,* 687–689.
Pandey, K. C. (1981), *Phys. Rev. Lett. 47,* 1913.
Pandey, K. C. (1992), *Phys. Rev. Lett. 49,* 223.
Pelz, J. P., Koch, R. H. (1990), *Phys. Rev. B42,* 3761–3764.
Persson, B. N. J., Baratoff, A. (1987), *Phys. Rev. Lett. 59,* 339–342.
Persson, B. N. J., Demuth, J. E. (1986), *Solid State Commun. 57,* 769–772.
Pohl, D. W., Denk, W., Lanz, M. (1984), *Appl. Phys. Lett. 44,* 651–653.
Prietsch, M., Ludeke, R. (1991), *Phys. Rev. Lett. 66,* 2511–2514.
Prietsch, M., Samsavar, A., Ludeke, R. (1991), *Phys. Rev. B43,* 11850–11856.
Qin, X.-R., Kirczenow, G. (1990), *Phys. Rev. B41,* 4976–4985.
Quate, C. F. (1991), *Nature 352,* 571.
Reddick, R. C. R., Warmack, R. J., Chilcott, D. W., Sharp, S. L., Ferrell, T. L. (1990), *Rev. Sci. Instrum. 61,* 3669–3677.
Reiss, G., Schneider, F., Vancea, J., Hoffmann, H. (1990), *Appl. Phys. Lett. 57,* 867–869.
Renner, C., Kent, A. D., Niedermann, P., Fischer, Ø., Lévy, F. (1991), *Phys. Rev. Lett. 67,* 1650–1652.
Rohrer, G. S., Bonnell, D. A. (1990), *J. Am. Ceram. Soc. 73,* 3026–3032.
Rohrer, G. S., Bonnell, D. A. (1991), *J. Vac. Sci. Technol. B9,* 783–788.
Rohrer, G. S., Henrich, V. E., Bonnell, D. A. (1990), *Science 250,* 1239–1241.
Rousset, S., Gauthier, S., Siboulet, O., Sacks, W., Belin, M., Klein, J. (1989), *Phys. Rev. Lett. 63,* 1265–1268.
Rudd, G., Novak, D., Saulys, D., Bartynski, R. A., Garofalini, S., Ramanujachary, K. V., Greenblatt, M., Garfunkel, E. (1991), *J. Vac. Sci. Technol. B9,* 909–913.
Rugar, D., Hansma, P. (1990), *Phys. Today 43,* 23–30.
Sakurai, T., Hashizume, T., Kamiya, I., Hasegawa, Y., Sano, N., Pickering, H. W., Sakai, A. (1990), *Progr. Surf. Sci. 33,* 3–89.
Salmeron, M., Beebe, T., Odriozola, J., Wilson, T., Siekhaus, W. (1990), *J. Vac. Sci. Technol. A8,* 635–641.
Samsavar, A., Hirchorn, E. S., Miller, T., Leibsle, F. M., Eades, J. A., Chiang, T. (1990), *Phys. Rev. Lett. 65,* 1607–1610.
Sarid, D., Elings, V. (1991), *J. Vac. Sci. Technol. B9,* 431–437.
Sass, J. K., Gimzewski, J. K., Haiss, W., Besocke, K. H., Lackey, D. (1991), *J. Phys.: Condensed Matter 3,* S121–S126.
Saurenbach, F., Terris, B. D. (1990), *Appl. Phys. Lett. 56,* 1703–1705.
Schmid, M., Varga, P. (1992), *Ultramicroscopy 42–44,* 1610–1615.
Schneir, J., Sonnenfeld, R., Hansma, P. K., Tersoff, J. (1986), *Phys. Rev. B34,* 4979–4984.
Schneir, J., Elings, V., Hansma, P. K. (1988), *J. Electrochem. Soc. 135,* 2774–2777.
Schottky, W. (1938), *Naturwissenschaften 26,* 843.
Schottky, W. (1950), *Phys. Z. 41,* 570.
Schroer, P. J., Becker, J. (1986), *IBM J. Res. Dev. 30,* 543–552.
Selci, S., Cricenti, A., Felici, A. C., Generosi R., Gori, E., Djaczenko, W., Chiarotti, G. (1990), *J. Vac. Sci. Technol. A8,* 642–644.
Selloni, A., Carnevali, P., Tosatti, E., Chen, C. D. (1985), *Phys. Rev. B31,* 2602–2605.
Selloni, A., Chen, C. D., Tosatti, E. (1988), *Phys. Scr. 38,* 297.
Shih, C. K., Feenstra, R. M., Chandrashekhar, G. V. (1991), *Phys. Rev. B43,* 7913–7922.
Silver, R. M., Ehrichs, E. E., de Lozanne, A. L. (1987), *Appl. Phys. Lett. 51,* 247–249.
Sleator, T., Tycko, R. (1988), *Phys. Rev. Lett. 60,* 1418–1421.
Slough, C. G., Giambattista, B., Johnson, A., McNairy, W. W., Coleman, R. V. (1989), *Phys. Rev. B39,* 5496–5499.
Slough, C. G., McNairy, W. W., Coleman, R. V., Garneas, J., Prater, C. B., Hansma, P. K. (1990), *Phys. Rev. B42,* 9255–9258.
Slough, C. G., McNairy, W. W., Chen Wang, Coleman, R. V. (1991), *J. Vac. Sci. Technol. B9,* 1036–1038.
Snyder, C. W., de Lozanne, A. L. (1988), *Rev. Sci. Instrum. 59,* 541–544.
Snyder, E. J., Anderson, M. S., Tong, W. M., Williams, R. S., Anz, S. J., Alvarez, M. M., Rubin, Y., Diederich, F. N., Whetten, R. L. (1991), *Science 253,* 171–173.
Soler, J. M., Baró, A. M., García, N., Rohrer, H. (1986), *Phys. Rev. Lett. 57,* 444–447.
Sonnenfeld, R., Hansma, P. K. (1986), *Science 232,* 211–213.
Sonnenfeld, R., Schardt, B. C. (1986), *Appl. Phys. Lett. 49,* 1172–1174.
Sonnenfeld, R., Schneir, J., Drake, B., Hansma, P. K., Aspnes, D. E. (1987), *Appl. Phys. Lett. 50,* 1742–1744.
Sotobayashi, H. (1990), *Langmuir 6,* 1246–1250.
Stern, J. E., Terris, B. D., Mamin, H. J., Rugar, D. (1988), *Appl. Phys. Lett. 53,* 2717–2719.
Stich, I., Payne, M. C., King-Smith, R. D., Lin, J.-S., Clarke, L. J. (1992), *Phys. Rev. Lett. 68,* 1351–1354.
Stoll, E. (1984), *Surf. Sci. Lett. 143,* L411–L416.
Stroscio, J. A., Feenstra, R. M., Fein, A. P. (1986), *Phys. Rev. Lett. 57,* 2579–2582.
Stroscio, J. A., Feenstra, R. M., Fein, A. P. (1987), *Phys. Rev. Lett. 58,* 1668–1671.
Suzuki, M., Homma, Y., Kudoh, Y., Kaneko, R. (1992), *Appl. Surf. Sci. 60/61,* 460–465.
Swartzentruber, B. S., Mo, Y., Kariotis, R., Lagally, M. G., Webb, M. B. (1990), *Phys. Rev. Lett. 65,* 1913–1916.

Tabata, T., Aruga, T., Murata, Y. (1987), *Surf. Sci. Lett. 179,* L63–L70.

Takayanagi, K., Tanishiro, Y., Takahashi, M., Takahashi, S. (1985), *J. Vac. Sci. Technol. A3,* 1502–1506.

Tanaka, S., Onchi, M., Nishijima, M. (1987), *Surf. Sci. Lett. 191,* L756–L764.

Teague, E. C. (1989), *J. Vac. Sci. Technol. B7,* 1989–1902.

Tekman, E. (1990), *Phys. Rev. B42,* 1860–1863.

Terris, B. D., Stern, J. E., Rugar, D., Mamin, H. J. (1989), *Phys. Rev. Lett. 63,* 2669–2672.

Terris, B. D., Stern, J. E., Rugar, D., Mamin, H. J. (1990), *J. Vac. Sci. Technol. A8,* 374–377.

Tersoff, J. (1984), *Phys. Rev. Lett. 52,* 465–468.

Tersoff, J. (1986), *Phys. Rev. Lett. 57,* 440–443.

Tersoff, J., Hamann, D. R. (1983), *Phys. Rev. Lett. 50,* 1998–2001.

Thibaudau, F., Dumas, P., Mathiez, P., Humbert, A., Satti, D., Salvan, F. (1989), *Surf. Sci. 211/212,* 148–155.

Thompson, W. A., Hanrahan, S. F. (1976), *Rev. Sci. Instrum. 47,* 1303–1304.

Thomson, R. E., Walter, U., Ganz, E., Clarke, J., Zettl, A., Rauch, P., DiSalvo, F. J. (1988), *Phys. Rev. B38,* 10734–10743.

Thundat, T., Nagahara, L. A., Oden, P., Lindsay, S. M. (1990), *J. Vac. Sci. Technol. A8,* 645–647.

Tománek, D., Louie, S. G., Mamin, H. J., Abraham, D. W., Thomson, R. E., Ganz, E., Clarke, J. (1987), *Phys. Rev. B35,* 7790–7793.

Tosch, S., Neddermeyer, H. (1988), *Phys. Rev. Lett. 61,* 349–352.

Tromp, R. M. (1989), *J. Phys.: Condensed Matter 1,* 10211–10228.

Tung, R. T. (1984), *Phys. Rev. Lett. 52,* 461–464.

Umeda, N., Ishizaki, S., Uwai, H. (1991), *J. Vac. Sci. Technol. B9,* 1318–1322.

Unertl, W. N., Jin, X., White, R. C. (1991), in: *Polyimides and other High-Temperature Polymers.* Amsterdam: Elsevier, pp. 427–435.

van der Walle, G. F. A., van Kempen, H., Wyder, P., Davidsson, P. (1987), *Appl. Phys. Lett. 50,* 22–24.

van Loenen, E. J., Demuth, J. E., Tromp, R. M., Hamers, R. J. (1987), *Phys. Rev. Lett. 58,* 373–376.

Villarrubia, J. S., Boland, J. J. (1989), *Phys. Rev. Lett. 63,* 306–309.

Wan, K. J., Lin, X. F., Nogami, J. (1992a), *Phys. Rev. B46,* 13635–13638.

Wan, K. J., Lin, X. F., Nogami, J. (1992b), *Phys. Rev. B45,* 9509–9512.

Wang, C., Giambattista, B., Slough, C. G., Coleman, R. V., Subramanian, M. A. (1990), *Phys. Rev. B42,* 8890–8906.

Watanabe, M. O., Tanaka, K., Sakai, A. (1990), *J. Vac. Sci. Technol. A8,* 327–329.

Watanabe, S., Aono, M., Tsukada, M. (1991), *Phys. Rev. B44,* 8330–8333.

Weaver, J. M. R., Walpita, L. M., Wickramasinghe, H. K. (1989), *Nature 342,* 783–785.

Weisenhorn, A. L., MacDougall, J. E., Gould, S. A. C., Cox, S. D., Wise, W. S., Massie, J., Maivald, P., Elings, V. B., Stucky, G. D., Hansma, P. K. (1990), *Science 247,* 1330–1333.

Weisenhorn, A. L., Henriksen, P. N., Chu, H. T., Ramsier, R. D., Reneker, D. H. (1991), *J. Vac. Sci. Technol. B9,* 1333–1335.

Weitering, H. H., Perez, R. (1993), unpublished.

Weitering, H. H., Chen, J., DiNardo, N. J., Plummer, E. W. (1993a), *J. Vac. Sci. Technol. A11,* 2049–2053.

Weitering, H. H., DiNardo, N. J., Perez-Sandoz, R., Chen, J., Mele, E. J. (1993b), unpublished.

Whitman, L. J., Stroscio, J. A., Dragoset, R. A., Celotta, R. J. (1991a), *Phys. Rev. Lett. 66,* 1338–1341.

Whitman, L. J., Stroscio, J. A., Dragoset, R. A., Celotta, R. J. (1991b), unpublished results.

Whitman, L. J., Stroscio, J. A., Dragoset, R. A., Celotta, R. J. (1991c), *Science, 251,* 1206–1210.

Wickramasinghe, H. K. (1990), *J. Vac. Sci. Technol. A8,* 363–368.

Wierenga, P. E., Kubby, J. A., Griffith, J. E. (1987), *Phys. Rev. Lett. 59,* 2169–2172.

Wiesendanger, R., Eng, L., Hidber, H. R., Oelhafen, P., Rosenthaler, L., Stauffer, U., Güntherodt, H.-J. (1987), *Surf. Sci. 189/190,* 24–28.

Wiesendanger, R., Güntherodt, H., Güntherodt, G., Gambino, R. J., Ruf, R. (1990), *Phys. Rev. Lett. 65,* 247–250.

Wiesendanger, R., Shvets, I. V., Bürgler, D., Tarrach, G., Güntherodt, H., Coey, J. M. D., Gräser, S. (1992), *Science 255,* 583–586.

Wilkins, R., Amman, M., Soltis, R. W., Ben-Jacob, E., Jaklevich, R. C. (1990), *Phys. Rev. B41,* 8904–8911.

Williams, C. C., Wickramasinghe, H. J. (1986), in: *IEEE – 1986 Ultrasonics Symp.* Piscataway, NJ: IEEE, pp. 393–397.

Williams, C. C., Wickramasinghe, H. J. (1991), *J. Vac. Sci. Technol. B9,* 537–540.

Wilson, R. J., Chiang, S. (1987a), *Phys. Rev. Lett. 59,* 2329–2332.

Wilson, R. J., Chiang, S. (1987b), *Phys. Rev. Lett. 58,* 369–372.

Wintterlin, J., Wiechers, J., Brune, H., Gritsch, T., Höfer, H., Behm, R. J. (1989), *Phys. Rev. Lett. 62,* 59–62.

Wolf, J. F., Vicenzi, B., Ibach, H. (1991), *Surf. Sci. 249,* 233–236.

Wolkow, R., Avouris, P. (1988), *Phys. Rev. Lett. 60,* 1049–1052.

Wolkow, R. A. (1992), *Phys. Rev. Lett. 68,* 2636–2639.

Wu, X. L., Lieber, C. M. (1990a), *Phys. Rev. Lett. 64,* 1150–1153.

Wu, X. L., Lieber, C. M. (1990b), *Phys. Rev. B41,* 1239–1242.

Wu, X.-L., Zhou, P., Lieber, C. M. (1988), *Phys. Rev. Lett. 61,* 2604–2607.

Young, R., Ward, J., Scire, F. (1972), *Rev. Sci. Instrum. 43,* 999–1011.

Youngquist, M. G., Driscoll, R. J., Coley, T. R., Goddard, W. A., Baldeschwieler, J. D. (1991), *J. Vac. Sci. Technol. B9,* 1304–1308.

Zasadzinksi, J. A. N., Helm, C. A., Longo, M. L., Weisenhorn, A. L., Gould, S. A. C., Hansma, P. K. (1991), *Biophys. J59,* 755–760.

Zheng, N. J., Tsong, I. S. T. (1990), *Phys. Rev. B41,* 2671–2677.

General Reading

Review Articles

Binnig, G., Rohrer, H. (1987), "Scanning Tunneling Microscopy – From Birth to Adolescence" (Lecture on the occasion of the presentation of the 1986 Nobel Prize in Physics). *Rev. Mod Phys. 59,* 615.

Güntherodt, H.-J., Wiesendanger, R. (1992), *Scanning Tunneling Microscopy I – General Principles and Applications to Clean and Adsorbate-Covered Surfaces.* Springer Series in Surface Sciences, Vol. 20. Berlin: Springer.

Hamers, R. J. (1989), "Atomic Resolution Surface Spectroscopy with the Scanning Tunneling Microscope". *Annu. Rev. Phys. Chem. 40,* 531.

Hansma, P. K., Tersoff, J. (1987), "Scanning Tunneling Microscopy". *J. Appl. Phys. 61,* R1.

Hansma, P. K., Elings, V. B., Marti, O., Bracker, C. E. (1988), "Scanning Tunneling Microscopy and Atomic Force Microscopy: Application to Biology and Technology". *Science 242,* 209.

Pool, R. (1990), "The Children of the STM". *Science 247,* 634.

Rugar, D., Hansma, P. (1990), "Atomic Force Microscopy". *Phys. Today 43 (10),* 23.

Wiesendanger, R., Güntherodt, H.-J. (1992), *Scanning Tunneling Microscopy II – Further Appplications and Related Scanning Techniques,* Springer Series in Surface Sciences, Vol. 28. Berlin: Springer.

Conference Proceedings

STM '86: 1st Int. Conf. on Scanning Tunneling Microscopy. Surf. Sci. 181, No. 1/2 (Mar 1987).

2nd Int. Conf. on Scanning Tunneling Microscopy. J. Vac. Sci. Technol. A6, No. 2 (Mar/Apr 1988).

4th Int. Conf. on Scanning Tunneling Microscopy. J. Vac. Sci. Technol. A8, No. 1 (Jan/Feb 1990).

5th Int. Conf. on Scanning Tunneling Microscopy/Spectroscopy. J. Vac. Sci. Technol. B9, No. 2, Part II (Mar/Apr 1991)

Ten Years of STM: 6th Int. Conf. on Scanning Tunneling Microscopy. Ultramicroscopy 52–44, Part A (July 1992)

Manufacturers

Information may also be available from the following manufacturers.

Angstrom Technology, Mesa, AZ.
Burleigh Instruments Inc., E. Norwalk, CT.
Digital Instruments, Santa Barbara, CA.
JEOL, Coldwater, MI.
LK Technologies, Inc. Bloomington, IN.
McAllister Technical Services, Coeur d'Alene, ID.
Omicron Associates, Pittsburgh, PA.
QuanScan Inc., Pasadena, CA.
Quantum Vision Corp., West Vancouver, B.C., Canada.
Park Scientific Instruments, Sunnyvale, CA.
RHK Technology Inc. Rochester Hills, MI.
Streuers Inc., Westlake, OH.
VG Instruments Inc., Danvers, MA.
WA Technology, Cambridge, U.K.

Index